Nuclear Power:
Villain or Victim?

OUR MOST MISUNDERSTOOD
SOURCE OF ELECTRICITY

Max W. Carbon

Emeritus Professor of Nuclear Engineering
University of Wisconsin-Madison

Third Printing

ISBN 0-9658096-0-9

Library of Congress Catalog Card Number: 97-91918

Title suggested by Karen Schrock.

Published by Pebble Beach Publishers
914 Pebble Beach Drive
Madison WI 53717

To my wife, Phyllis

FOREWORD

Oh, the things we think we know
that are not so.
Anonymous

Making electricity from nuclear energy is a fascinating process. How in the world can we generate electricity to operate a microwave oven from the nuclei of atoms — specks of matter so tiny that 10,000 billion of them would still be invisible? This book is about nuclear power — electricity made from specks of uranium and plutonium.

Nuclear power is important to Americans for jobs, a high standard of living, and clean air. Because it emits no pollutants to the atmosphere, it is estimated to save thousands of lives every year in the United States alone; it could save tens of thousands more. It may be crucial for preventing catastrophic consequences of global warming and for preventing wars over the world's supply of petroleum.

Polls indicate that a majority of Americans support nuclear power, but a vocal minority has opposed it. That minority includes people who are sincere and well-meaning; people who do not understand it; environmental groups, some of which use opposition to it as a fund-raising tool; and antinuclear groups that oppose both advanced technology and large industry in general.

A lack of understanding of nuclear power is a major cause of opposition. Some people believe radiation from nuclear energy is new and man-made, although the earth has been bathed in it since the dawn of time. The public has feared radiation since the two atomic bombings in Japan in 1945; however, although the radiation effects were severe, radiation accounted for only 20% of the deaths there. The mass media contribute to misunderstanding through the use of frightening headlines, frequently to attract readers or listeners. A New York newspaper reported "over 20,000 dead" a few days after the Chernobyl nuclear power plant failure in Ukraine in 1986, although the number of known deaths is 34. Individuals make outrageous statements. An antinuclear activist has said that one pound of plutonium could kill eight billion people; however, 10,000 pounds have been released into the earth's atmosphere from weapons tests in the last 40 years — enough by his estimate to kill all the people on earth several thousand times.

There is also misunderstanding about nuclear waste. Antinuclear groups and some political leaders state repeatedly that the nuclear waste disposal problem is unsolved, and the public comes to believe this. However, most of our scientific and engineering societies believe the waste can readily be disposed of by deep-underground burial — where it will be harmless. Even if the waste-disposal statement were true, it could be quite misleading; it is intended to imply that other technologies do not have significant waste-disposal problems. The groups rarely mention that there is no solution for handling the several million tons of carbon dioxide that every large coal or natural gas power plant discharges each year — other than to release the gas to the atmosphere where it becomes a major contributor to the greenhouse effect and potential climate change. Nor do we have an easy solution for handling particulate pollutants that coal plants discharge into the atmosphere — where they are estimated to cause tens of thousands of deaths yearly.

This book's major goal is to present facts about nuclear power and to eliminate as much misunderstanding about it as possible. The book is addressed to adults who have forgotten their high school science courses and to ninth and tenth grade high school students who haven't; brighter students in grades as early as the fourth or fifth will also understand it. Each of you will learn more about the subject than most scientists, engineers, government leaders, and representatives of industry and environmental groups know. The book doesn't provide the intricate details of the design and operation of a nuclear plant because that takes years of study. Nevertheless, it gives "the big picture" on which decisions are made about the use of nuclear energy.

We will compare nuclear power with its alternatives, just as everybody compares alternatives in real-life situations. If you intended to buy a car, you would compare prices, styles, gas mileages, trade-in values, and colors of different makes and models. You wouldn't simply walk to the nearest automobile dealer and buy the first car you saw in the window. Similarly, we must make comparisons in deciding the best ways to make electricity. The primary alternatives are to make it from burning fossil fuels — coal, natural gas, and oil (petroleum). Of course, we first have to decide whether we want electricity.

I will present considerable data for making comparisons. This is partly because most readers don't have time to find the needed information by themselves — the data have to be dug out from many books and technical magazines. A second reason is that those books and magazines are not easily available for most people. Some data will surprise (shock?) you, and you may doubt the accuracy of some statements. Good. You are urged to check all the information presented, and a Suggested Reading list is given at the end of the book to help. Your public or school libraries should be able to obtain these references.

Chapters 1 through 10 cover the following:

- What is nuclear power and why is it important?
- What is nuclear energy?
- What is a nuclear reactor and how does it work?
- How is electricity made from nuclear energy?
- What are the health effects of radiation?
- Are nuclear power plants safe?
- How do we dispose of radioactive waste?
- What is the possibility of theft of uranium or plutonium by terrorist groups to make explosives?
- What kind of advanced reactors are being developed?
- What is the cost of nuclear power?

Chapter 11 describes the enormous benefits that nuclear energy promises the world. Chapter 12 gives recommendations on what we can do to help realize those benefits.

ACKNOWLEDGMENTS

I am very grateful to several people who have offered advice and help in preparing this book. They include Seymour Abrahamson, Professor Emeritus of Genetics and Zoology, University of Wisconsin-Madison; John R. Cameron, Professor Emeritus of Radiology, University of Wisconsin-Madison; Mr. Ronald A. Carbon, Marketing Executive, Acxiom Corporation; Dr. Melvin S. Coops, Staff Chemist-Retired, Lawrence Livermore National Laboratory (LLNL); William Kerr, Professor Emeritus of Nuclear Engineering, University of Michigan; Mr. Charles A. Schrock, Plant Manager, Kewaunee Nuclear Power Plant; Ms. Karen Schrock, a Sophomore at Southwest High School in Green Bay, Wisconsin; Dr. George Sliter, Nuclear Power Group, Electric Power Research Institute; and Dr. Carl E. Walter, Engineer-at-Large, Emeritus, LLNL. I am especially indebted to Professor Cameron, who urged me for two decades to write the book and then showed me how to do it in understandable language. However, because I made the final decisions on wording and content throughout the text, I accept full responsibility for any errors.

CONTENTS

Chapter 1

WHY IS NUCLEAR POWER IMPORTANT?

There shouldn't be any mystery about what nuclear power is. It is simply electricity produced from nuclear energy using a nuclear reactor, a turbine, and an electrical generator. There is nuclear energy in uranium, and we can release this energy in a reactor. We can then convert the energy to heat and use the heat to boil water and make steam. Finally, we can cause the steam to turn the turbine, which is attached to a coil of copper wires located inside a magnet. We call this the generator. If we do everything properly, electricity is produced as the wires turn inside the magnet.

Now, a logical question is: Why is electricity important? One reason is that it helps maintain and increase our standard of living as our population grows. Another reason is that it allows our factories to increase their efficiency. Low-cost electricity is vital for industry to compete internationally and to provide jobs. Our electricity use has risen continuously for the last 40 years, and it is expected to grow about 1.5% per year through 2015. This growth and the replacement of older power plants as they wear out will require construction of over 300 large new plants by that date — 18 years from now; they will cost many tens of billions of dollars.

The need for electricity is even greater outside the United States. Today, there are about six billion people in the world, and almost two billion of them have no electricity. In 50 years, there will likely be about 10 billion people on the planet.

A second question is: Why is nuclear energy important for making electricity? The answer is that it is a clean, safe, and inexpensive way to produce electricity. Nuclear energy is especially important for *clean air*. Most electricity produced in the world today comes from burning coal, natural gas, and oil, including about 70% of that in the United States. Burning these fossil fuels releases carbon dioxide (CO_2) into the atmosphere. It is estimated that two billion tons of CO_2 per year are released into our air from generating electricity. China releases 50% more than the United States does, and its burning of coal may double in 25 years. CO_2 is responsible for about two-thirds of the "greenhouse effect" and potential global warming; many scientists predict this warming will cause disastrous climate changes in parts of the world. CO_2 is released when any fossil fuel is burned, including natural gas that releases about 55% as much as coal in making electricity. In contrast, no CO_2 is released when uranium is used to produce electricity. Coal plants also release sulfur and nitrogen compounds that cause acid rain, whereas nuclear plants do not.

Nuclear energy is also important to *save lives*. Nuclear power is safe; except for the Chernobyl event in Ukraine in 1986, there have been no known deaths among the public caused by the world's nuclear plants during their 36 years of operation. Even Chernobyl will not cause more than a few hundred deaths by optimistic estimates or in the range of 24,000 by pessimistic ones. In contrast, the burning of fossil fuels, such as coal, diesel oil, and gasoline, releases tiny particles or particulates into the atmosphere; breathing these particulates is estimated to cause tens of thousands of deaths each year in the United States alone. Nuclear plants do not release particulates. Because 20% of our electricity comes from nuclear power, the 100-plus nuclear plants operating in the United States today are probably saving thousands of lives every year. Increased substitution of nuclear power for electricity from burning coal would save thousands more lives each year.

The only proven methods to generate large amounts of electricity at *competitive costs* in new plants are to burn fossil fuels or to use uranium. Our American dam sites for hydroelectric power have largely been put to use, as have our geothermal sites. Electricity generation from solar energy, wind, wood, or other "renewable" sources has not been demonstrated on a large scale and appears to be considerably more expensive than from coal or uranium. Power from controlled fusion is at least 40 years away. In contrast, the cost of electricity from uranium in new plants is expected to be only a few percent higher than the costs from natural gas and coal, and less than from oil. In France, electricity from uranium is 25% cheaper than from coal. Further, the nuclear energy in a pound of uranium is three million times the energy released in burning a pound of coal; the long-range potential cost of generating electricity from uranium is considerably lower than from any fossil fuel.

> Note: Not all nuclear power plants produce electricity. Navy nuclear plants propel ships directly without using generators or electricity. However, we will continue to discuss nuclear power as electricity produced with reactors. Electricity can also be produced on a small scale from nuclear energy without a reactor. This is done on space satellites, using a different kind of nuclear process.

Chapter 2

WHAT IS NUCLEAR ENERGY?

To understand nuclear power, we need to understand what nuclear energy is and where it comes from. According to Einstein's famous equation, $E = mc^2$, mass (m) can be converted into energy (E) in a nuclear process. (In the equation, c is the speed of light.) Therefore, the more mass we use up, the more energy we get. It is easy to convert uranium into energy, and we do it as follows.

As you may recall, there are 91 elements that occur in nature, and almost everything in the world is made up of those individual elements or combinations of them. When we study the elements in science or chemistry, they are listed in a periodic table, with hydrogen as number one, helium as number two, and so on until we reach 92, which is uranium. (The 43rd element, technetium, does not occur naturally.) If you held a chunk of uranium in your hand and were able to crumble it into very tiny pieces, you would find that it is made up of billions of individual particles called "atoms." The atoms for each element have a nucleus at the center and a unique number of electrons outside the nucleus; hydrogen, helium, and uranium atoms are illustrated in Fig. 1 through 3.

Electrons are tiny bits of electricity, and if we try to bring two of them together, we would find that they repel one another. Protons are much bigger (but still tiny) particles, and they, too, repel one another. However, if we bring an electron and a proton close together, we would find that they attract each other; it would be difficult to keep them apart. Neutrons are similar in size to protons, but they neither repel nor attract one another or electrons or protons.

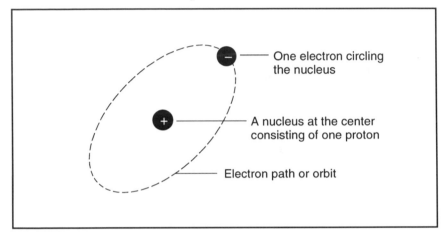

Figure 1: Hydrogen Atom

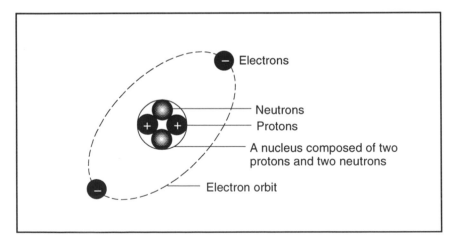

Figure 2: Helium Atom

No one has ever seen an atom, an electron, a proton, or a neutron, nor has anyone counted the number of protons or neutrons in a nucleus. They are all too small. However, scientists (particularly physicists and chemists) have been studying atoms for over 200 years and have developed models to explain the results of the thousands of experiments they have run.

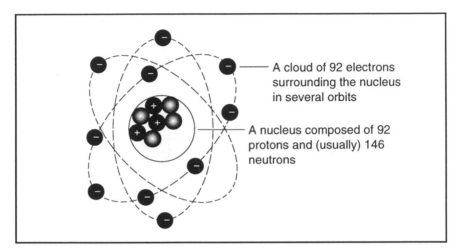

A cloud of 92 electrons surrounding the nucleus in several orbits

A nucleus composed of 92 protons and (usually) 146 neutrons

Figure 3: Uranium Atom

Our model tells us that uranium nuclei are unstable and some are continually breaking up or disintegrating and emitting neutrons. Further, the model and experiments show that, if one of these neutrons hits and is absorbed in the nucleus of another uranium atom, the nucleus may split into two fragments; it will also release two or three neutrons, and give off energy. We term this splitting process "fissioning," and the energy is termed "nuclear" energy because it comes from a reaction in the nucleus. The model is illustrated in Fig. 4.

Let us examine the energy further. First, the energy results from some of the mass in the nucleus being converted to energy as in Einstein's equation; if we could weigh the fission fragments and neutrons that result from splitting the nucleus, we would find that they weigh less than the original nucleus and neutron. The difference in weight has gone into energy.

(For the purposes in this book, we will use mass and weight interchangeably, although that is a simplification.)

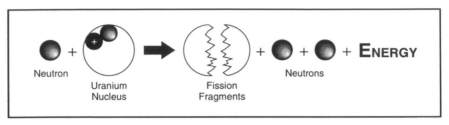

Figure 4: Fission Process

We say there are two kinds of energy — kinetic and potential. Kinetic energy is the energy of motion — the energy of a body or object that results from its motion. A moving automobile and a bullet fired from a gun have kinetic energy. Potential energy is energy that can be converted to kinetic energy, and it results from location or structure. A rock sitting on top of a hill has potential energy; it could roll down the hill and its potential energy would be converted to kinetic energy.

Kinetic energy can be converted easily to heat. If I swing a heavy hammer (which has kinetic energy because it is moving) and strike a piece of iron, the hammer will lose part of its energy, the iron will absorb that energy, and the iron will rise in temperature. A similar happening occurs if you rub one stick against another to start a fire — kinetic energy is converted to heat, which raises the temperature of the wood.

The energy given off when a uranium nucleus absorbs a neutron and fissions is kinetic energy; the two fission fragments each travel at a high speed. They will be slowed down and stopped as they strike surrounding atoms, and the piece of uranium will be heated.

Nuclei are very tiny, as discussed earlier, and the amount of energy given off in an individual fission event is extremely small. However, huge numbers of fission events (typically 10,000,000,000,000,000,000) can be made to occur every second in a nuclear reactor power plant, and large amounts of energy can be released.

One important aspect of fissioning uranium has not been mentioned so far. There are several kinds of uranium atoms, each called an *isotope* of uranium. The isotope pictured in Fig. 3 with 92 protons and 146 neutrons in the nucleus is called uranium 238 or U-238. (Note that 92 + 146 = 238.) If you were to dig up a piece of uranium ore in Colorado, you would find that about 99.3% of the uranium atoms would be U-238. Most of the remaining 0.7% would be uranium 235 atoms, with 92 protons but only 143 neutrons in each nucleus. There is an important difference between the two isotopes: it is easy to cause a U-235 nucleus to capture a neutron, to fission, and to release energy, whereas this is difficult with U-238. In the electricity-generating reactors used around the world today, little fissioning takes place in U-238 nuclei, and we will ignore it in this book.

U-238 nuclei are important though because they can absorb neutrons to form the element plutonium, which does not occur in nature. Most nuclei of plutonium will fission and release energy just like U-235. Plutonium is produced in commercial power reactors during normal operation and about 40% of the electricity comes from it.

In summary, the term "nuclear energy" as used in this book is the kinetic energy of the fragments that result from the fissioning or splitting of U-235 and plutonium nuclei when they absorb neutrons. The kinetic energy is converted to heat as the fragments are slowed, and the heat is converted to electricity as described later.

In passing, it should be noted that nuclear energy can also be released by fissioning in a process involving thorium, the 90th element in the periodic table. However, thorium is seldom used and will not be discussed further in this book.

Actually, there are two processes by which nuclear energy is released — fission and fusion. The energy given off by our sun is another form of nuclear energy; it comes from the fusion process. In this, the nuclei of atoms combine or fuse together, and mass is lost in the process. The mass is converted into energy, also in accordance with Einstein's equation. Fusion is much more difficult to achieve here on earth than fission; this is partly because it occurs at temperatures around 100 million degrees. Scientists and engineers have been working since 1950 to learn how to make power from fusion, but they have not yet been successful. We probably will not have fusion electricity for at least 40 years, and it will not be discussed further.

Chapter 3

WHAT IS A NUCLEAR REACTOR?

What is a "nuclear reactor"? It is a machine we build to release energy from uranium and plutonium. The energy is then converted to heat and used to heat water[*] and make steam. (The steam is used to make electricity as described later.) In some reactors, the steam is made inside the reactor; in others, very hot water is piped outside the reactor to form steam there.

A typical reactor consists of four main parts: uranium or uranium and plutonium; water; devices to control the rate at which fission occurs; and a radiation shield, which is discussed in Chapter 6. The water is used to (a) cool the uranium, (b) make steam, and (c) slow down the neutrons as follows.

Suppose we take a block of pure uranium metal. As discussed previously, we would find that some nuclei are continuously breaking up, with neutrons being released in the process. These neutrons would fly off in all directions at very

---

[*] Helium, carbon dioxide, a type of water called heavy water, and the metal sodium (in liquid form) are substituted individually for ordinary water in some reactors. In particular, Canadian reactors use heavy water and some British reactors use carbon dioxide. Sodium reactors are discussed later.

high speed. Some would strike nearby U-235 nuclei and cause fissioning to take place. However, this is an inefficient process because the neutrons travel too fast (about 40 million miles per hour); many of them would escape from the block and be lost. To improve the efficiency, let us shape the uranium into a large number of small rods, each about one-half inch in diameter and several feet long. Let us also arrange the rods vertically and insert water between them to slow down the neutrons. The neutrons slow down (to perhaps 7,500 miles per hour) because they lose energy as they strike the nuclei of hydrogen in the water. They then more easily cause fission of U-235 nuclei. In a typical reactor, the rods are separated from each other by about one-eighth inch of water.

The two other purposes of the water are clear. First, it is pumped past the uranium rods to carry away the heat. Unless the uranium is cooled, it would melt. Second, the heated water forms steam to generate electricity as described later.

The fission rate must be controlled or the reactor could be destroyed. This is done by control rods. At least two neutrons are released in each fission event, and the number of neutrons in the reactor could multiply to an undesirably high level. In fact, it could rise so high that the water would be incapable of carrying the heat away and the uranium rods would melt. To prevent this, we insert an additional material into the reactor to absorb "excess" neutrons and control the fission rate. The element boron is frequently used because it absorbs neutrons readily. By moving the boron into or out of the uranium region, we can easily control the rate of fissioning. Of course, if we insert too much boron, the neutron population drops rapidly. This is the way the reactor is shut down.

A "boiling water reactor" and a "pressurized water reactor" are illustrated in Fig. 5 and 6. In the boiling water reactor, water enters the reactor and turns to steam as it passes through. In the pressurized water reactor, hot water under very high pressure leaves the reactor and is passed through tubes in a "heat exchanger"; here, heat from the reactor water passes through the walls of the tubes and boils a separate supply of water outside the tubes.

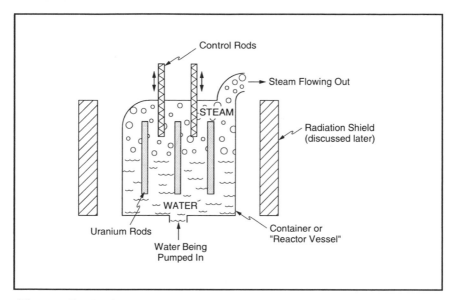

Figure 5: Boiling Water Reactor

And that is all a reactor is; it can be a very simple device. In fact, nature created one in what is now the African nation of Gabon about two billion years ago. There was enough uranium and water in the ground for fissioning to occur intermittently for several hundred thousand years. The water slowed the neutrons and served as a control device. With water present, the neutrons would be slowed and fissioning would occur; if the water boiled away, the reactor would shut down until rain replenished the supply. Then, fissioning would begin again. Of course, nature didn't care about collecting steam to generate electricity.

How do we know this happened? Because the fragments given off in fissioning of U-235 nuclei are effectively "fingerprints," and the fingerprint evidence at the site is overwhelming. In addition, the fingerprints can be dated to tell when the event occurred. Could such reactors have occurred elsewhere in the world? Yes, and they probably did.

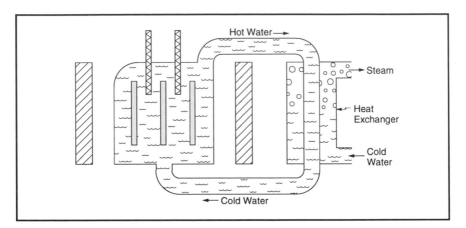

Figure 6: Pressurized Water Reactor

Chapter 4

WHAT IS A NUCLEAR POWER PLANT?

So far in our story, we have a reactor with steam being produced. How do we make a power plant — how do we generate electricity?

Let us discuss the last component of a power plant — the electrical generator. The discovery of how to build such a machine was made 150 years ago. If you wind copper wire into a circular coil and rotate the coil inside a magnet, you can generate electricity. Stationary brushes rub against the coil and carry the electricity away. A sketch of a generator is shown in Fig. 7.

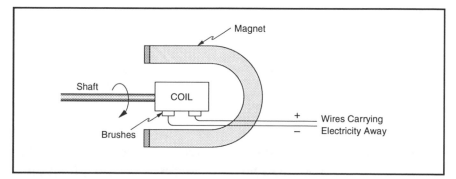

Figure 7: Electrical Generator

How can we couple the reactor to the generator to make electricity? The answer is by using a machine called a turbine. A turbine is like a windmill that you may have seen in a farmer's field. It is basically a big wheel with many "cups" at the edge; it turns when steam (or air in the case of the windmill) blows on the cups, as shown in Fig. 8. It has a steel shaft in the center of the wheel that is connected to the coil of copper wire in the electrical generator. Therefore, the steam turns the turbine wheel; the shaft at the center of the wheel turns the coil of wires; and electricity is produced when the coil turns inside the magnet. The wires forming the coil are connected to transmission lines via brushes that carry the electricity to homes and factories.

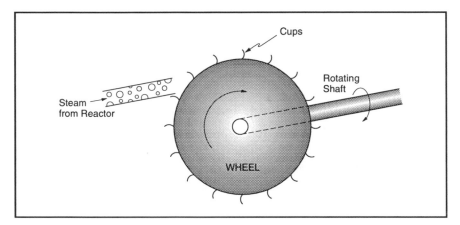

Figure 8: Steam Turbine

Now, the turbine wheel won't turn unless the steam from the reactor is flowing at a high speed. The final question to be answered, then, is why the steam from the reactor is flowing rapidly. Let's use a tea kettle to explain.

If you heat a tea kettle on a stove, the temperature of the water inside will rise to the boiling temperature. Steam will be formed, and the steam will push the spout open and "shoot out" or escape. This happens because, as the water turns to steam, its volume expands about 1,000 times. This expansion causes the pressure to build up in the kettle, and the pressure forces the steam out at high speed. If you held your finger on the spout so the steam couldn't escape, the pressure inside would build up and the kettle would rupture. (Don't try this; you might get hurt.)

A reactor behaves in a similar manner. As the reactor water is boiled, its volume increases, and the steam escapes at high speed through the outlet piping. The piping is designed so the steam strikes the cups on the turbine wheel; the wheel spins and its shaft turns the copper coil in the electrical generator.

A complete nuclear power plant is illustrated in Fig. 9.

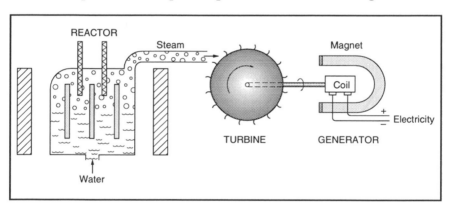

Figure 9: Boiling Water Reactor Nuclear Power Plant

There is one more component, in particular, that must be mentioned. We would like to reuse the steam after it leaves the turbine; very pure water is used in the reactor, and it is more economical to reuse the steam than to continuously purify replacement water. Liquid water can be pumped but steam can not. Consequently, we must condense the steam back to liquid water after it leaves the turbine; this necessitates removing heat from the steam. This can be accomplished in a "condenser" as shown in Fig. 10.

The Kewaunee and Point Beach Nuclear Power Plants in Wisconsin are located on Lake Michigan; they take water from the lake, use it for condensing steam, and return it to the lake. In the condenser, the lake water flows inside metal tubes and the steam flows on the outside of the tubes. Heat passes through the tube walls from the steam to the lake water. The steam condenses as it loses heat, and the lake water is warmed as it receives heat. Thus, the water is a few degrees warmer when it returns to the lake than when it left. Many fish like warm water and congregate at the condenser outlet; fishermen frequently have great success there. It should be recognized that the lake water is kept totally separated from the reactor water; the two supplies of water never mix, and the lake water never comes near the reactor.

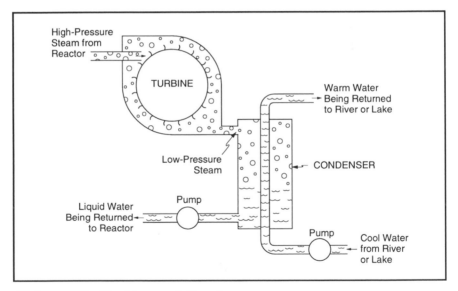

Figure 10: Turbine and Condenser

When power plants can not be located near lakes or large rivers, another method of providing cooling water for the condenser is required. "Cooling towers" are frequently used, and they substitute for a lake. This system is shown in Fig. 11.

Cold water is pumped through the condenser and is heated as the steam condenses. This warm water is then sprayed downward from the top of the cooling tower. At the same time, a big fan blows cool air upward inside the tower. As the air and water pass each other, the air cools the water. The air and a little water vapor are discharged to the atmosphere; it is the vapor or "steam" that you see from a distance. The cooled water is pumped back through the condenser. As Fig. 11 shows, the water is recirculated continuously through the condenser and the cooling tower. The water is heated in the condenser as it absorbs heat from the steam, and it is cooled by the air in the cooling tower. Fresh water must be added continuously to replace the water lost as vapor, of course. Note that this cooling tower water is entirely separate from the reactor water; as before, the cooling tower water never gets near the reactor. These towers are huge structures, many about 500 feet tall. They can be seen from a considerable distance from nuclear plants that use them.

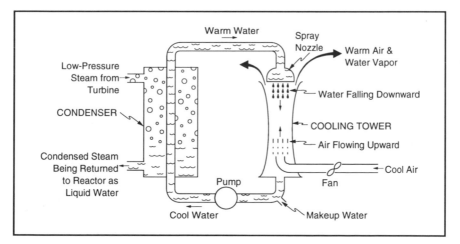

Figure 11: Cooling Tower

It is worth noting that about two-thirds of the nuclear energy released in the uranium and plutonium is lost through the condenser cooling water to the lake or atmosphere; the plant has a "thermal efficiency" of about 33%. Some advanced reactors show promise of achieving 50% efficiency. It is unfortunate that such a small fraction of the released energy is converted to electricity. However, this process is the most practical energy-conversion technique we have for large-scale electricity production.

A coal-burning power plant is somewhat more efficient, but it operates much the same way. Coal is currently the predominant fuel for electricity generation in the United States, followed by uranium. A coal plant is shown in Fig. 12.

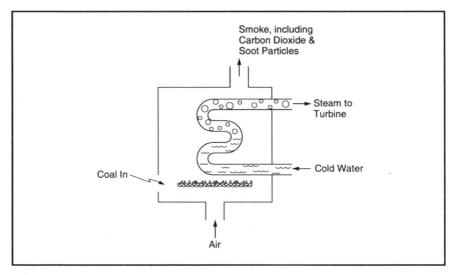

Figure 12: Coal Plant Furnace

When coal is burned, the carbon (C) in the coal combines with oxygen (O_2) from the air to form CO_2, and chemical energy is released. (This is truly "atomic" energy; atoms of carbon and oxygen combine chemically to form CO_2. The nuclei of these atoms are unchanged and play no role here.) The CO_2 is discharged to the atmosphere as part of the "smoke." The remainder of the system, including the turbine, electrical generator, condenser, and cooling tower, is basically the same as for the nuclear plant. Heat is also lost when the steam from the turbine is condensed. The thermal efficiency in a coal plant is usually somewhat higher than in a nuclear plant; efficiencies above 40% are typical in new coal plants.

Plants burning oil or natural gas are quite similar to coal plants, with carbon and oxygen again combining to release energy. As with coal, the CO_2 is discharged to the atmosphere.

Chapter 5

RADIATION AND HEALTH EFFECTS

What Is Radiation?

Although there are many similarities between nuclear and fossil plants, there are also several unique differences; one is that there are much larger amounts of radiation associated with nuclear than fossil units. We will explore radiation in this chapter.

Everyone has heard of radiation, but what is it? There are many kinds, but our discussion will center on four: the electrons and neutrons discussed in Chapter 2; the "alpha particle," which is the nucleus of the helium atom (two protons and two neutrons, as shown in Fig. 2); and a fourth kind, called a gamma ray. Gamma rays are tiny bundles or packets of energy that are weightless. They are very similar to X-rays and the bundles of energy that make up ordinary sunlight. Gamma rays and X-rays have more energy than a sunlight bundle. This extra energy makes them invisible to us. Thus, the radiation of importance in this book consists primarily of three particles (electrons, neutrons, and alpha particles) and little packets of energy (gamma rays).

Where Does Radiation Come from?

Radiation is natural; life evolved in a sea of radiation. It is in our bodies; our food; and in the soil, water, and air. The radiation level now is about one-tenth of what it was when life began billions of years ago.

There are two sources of radiation that are important to us in this discussion: nature and humans. From nature, each of us receives radiation termed "background radiation." It comes from the sun and outer space; from materials inside our bodies; from materials such as uranium and thorium in the soil and in our buildings; and from radon, a radioactive gas found in the soil which leaks into our homes. Everyone on earth receives this type of radiation. It is mostly alpha and gamma radiation. It is invisible, although it is very easy to detect and measure with instruments or even photographic film. The amount we each receive varies, depending primarily on where we live. Much of the radiation that strikes a person's body simply passes through without "touching" us; it has no effect on us. However, a small amount is absorbed and may have an effect. The terms used by scientists to express quantities of absorbed radiation are confusing. Therefore, we will discuss radiation in terms of the amount the typical American absorbs each year and call that amount *one year of background radiation*. For the reader who wishes to refer to other books, this amount of radiation is defined as 3 mSv (milliSievert) per year or 300 mrem (millirems) per year.

Man-made radiation comes primarily from medical X-rays. A small amount comes from various devices such as nuclear reactors and smoke detectors. Our interest will center on radiation produced in the reactors by fissioning uranium and plutonium nuclei. There are three categories. First, when a uranium or plutonium nucleus captures a neutron and fissions, electrons, gamma rays, and neutrons are emitted instantaneously. The second category comes from the fragments into which the uranium or the plutonium nuclei split. Nearly all of these fragments emit radiation. Some release their radiation almost immediately; others release theirs over a period of hundreds or thousands of years. Most of this radiation consists of electrons and gamma rays.

The third category of radiation related to fission comes from ordinary materials that absorb neutrons and subsequently emit radiation. For example, there is a moderately common metal called cobalt; it is important in making inks, paints, and stainless steel. If we put a piece of cobalt in a nuclear reactor, many of its nuclei will absorb neutrons. These nuclei will each eventually emit one electron and two gamma rays, and we say the cobalt has become *radioactive*. This is how the radioactive cobalt used in hospitals to treat cancer is made. Many materials can be made radioactive by exposing them to neutrons. The iron used in the structural supports inside reactors becomes radioactive. Most radioactive materials made this way will emit electrons and gamma rays, but some will emit neutrons.

What Are the Health Effects of Large Amounts of Radiation?

Much radiation is beneficial to us. Over 100 million Americans have a total of about a billion X-rays each year; 10 million Americans are diagnosed using radioactive medicine yearly; and a quarter million cancer patients are treated with radiation each year — many have their cancers cured. The medical use of radiation saves thousands of lives every year. Sunlight is radiation, which makes life on earth possible.

Some radiation has no detectable effect on us. For example, there is always radioactive potassium (element 19) in the food we eat, and some of this is stored in our bodies. About 18 million potassium nuclei disintegrate in the body of a typical adult and emit radiation each hour. The released radiation strikes billions of our cells every hour. It does no apparent harm. About an equal number of radioactive carbon nuclei emit electrons in our body.

However, many scientists believe that large amounts of radiation to prospective parents can genetically harm their children and grandchildren; large doses of radiation have been shown to produce harmful mutations in all plant and animal systems studied. Further, very large doses can cause sickness or death. Let us examine these effects.

About 70 years ago, scientists discovered that high doses of radiation damage both the chromosomes and their genes in fruit flies; this damage causes mutations and abnormal genetic effects in the offspring. The scientists assume that similar results will occur in humans. However, no such effects have been found. Our main source of information comes from studies of children of the survivors of atomic bomb blasts in World War II (1945). The bomb released at Hiroshima, Japan (made of U-235) and the one at Nagasaki, Japan (made of plutonium) derived their tremendous energy from nuclear fission. They emitted huge quantities of radiation. However, extensive studies of 30,000 children born to parents who were exposed to radiation in the blasts have found no evidence of genetic effects. This result is not surprising; the frequency of mutations is so low that they would likely be detectable only in a much larger group of children. In addition, the parents would have had to receive considerably more radiation than the bomb survivors.

The situation is different with regard to non-genetic effects; if the body absorbs very large amounts of radiation in a few minutes or hours, sickness or death can result. These are termed *acute* doses — doses received in a short time period. Our main source of information on these effects also comes from the 1945 explosions. About 400,000 civilians were present in Hiroshima and Nagasaki at the time of the explosions; 175,000 of them died instantly or within four months. Most of the deaths (80%) were due to blast and heat, but the other 20% (35,000) were caused by radiation, as stated earlier. Information has also come from medical people who use X-ray machines and from patients treated by radiation. There have also been a few accidents in which people received large, acute doses of radiation.

From all the evidence, we know that about half the people who absorb 1,500 times as much radiation in a few minutes as we normally do in a full year will die within a month. An acute dose equal to about 650 years of background radiation will cause sickness, but nearly all people will recover; there will be almost no deaths except from cancer as discussed below if proper medical care is received.

We also know that acute doses of radiation below 650 years of background or large doses received over an extended period of time will increase the chance of cancer. Information on the effects of acute doses comes from bomb survivors and is discussed in the next section below. Our primary information on radiation received over a long time period comes from studies of workers who used radium to make luminous watch dials; radium is radioactive and emits alphas, electrons, and gamma rays. From 1915 to about 1960, thousands of young women were hired in factories to paint radium solutions on dials, and they did this with tiny paintbrushes. Unfortunately, they sharpened the tips of the brushes by touching the brushes to their tongues. Radium entered their bodies, and about 2% (85 out of about 4,000) of the painters died from bone cancer many years later. No one who started work after 1925 died of radiation-induced cancer; beginning in that year, workers were forbidden from touching the brushes to their tongues.

What Are the Health Effects of Small Amounts of Radiation?

The amount of radiation that the public receives from nuclear power plant operation is *thousands of times below the levels discussed above* that cause sickness or death. It is also *several hundred times below background levels*. During normal operation, the amount of radiation leaving a plant site is so small it is almost unmeasurable. Releases during accidents are also minimal. There has been only one major accident in the United States in nuclear power's 36-year history — at Three Mile Island in Pennsylvania in 1979. No member of the public received more than the equivalent of one-third year of background radiation from it; at worst, not more than one person will die from it. Expressed differently — the average person living near the accident site received less radiation in 1979 than nearly every person in Denver did that same year. This was because the background level in Denver is higher than that in Pennsylvania. Denver has more uranium in its soil and is at a higher elevation than the TMI area, which allows more cosmic radiation. The people in Denver receive this high level of background radiation every year of their lives, of course.

What do we know about the health effects of small quantities of radiation? Fortunately, quite a bit, although not as much as we would like. First, no study has ever shown a harmful effect of radiation at doses below 65 times the amount we normally receive as background in a full year. Scientists at the *Radiation Effects Research Foundation* in Hiroshima, Japan have studied about 86,000 Japanese bomb survivors since 1950. They find that some survivors who received large doses of radiation developed cancer later in life. However, they have found no meaningful evidence of cancer formation among the 75,000 survivors who received 65 times yearly background or less. There were cancer deaths, but they were so few that the scientists couldn't tell whether any resulted from bomb radiation. In contrast, among a group of 6,308 people who received larger doses (between 65 and 165 times our normal yearly background), there were enough deaths that the scientists could conclude many were caused by bomb radiation. There were 659 deaths between 1950 and 1990, whereas only 567 deaths would have been expected if the people had not been exposed to bomb radiation. Thus, it appears that bomb radiation caused about 92 deaths; this is also evidence that harmful effects become observable somewhere in the range of 65 to 165 times yearly background.

I believe most of the scientific community would agree with the statements in the above paragraph. Beyond that point, however, there is sharp disagreement. One group, composed of genetic and cancer specialists, theorize that, if a large amount of radiation will cause 100 cancer deaths, half that amount will cause 50, one fourth the amount will cause 25, and so on. They carry this extrapolation down to zero radiation exposure and believe that any amount of radiation is harmful, no matter how small. They make no distinction between acute radiation and radiation received over a long time period. Their theory is called the *linear, no-threshold theory*. There is considerable evidence from bomb survivors that this theory is correct at high, acute radiation exposures — exposures more than 65 times normal background for a year. As stated earlier, there are no supporting data at lower levels. This is partly because, as the doses get smaller and smaller, the size of the group of people studied must become larger and larger to show an effect. However, the members of this group believe that extensive, related studies at the chromosome and gene level of cells support their beliefs. Their views are accepted by the National Council on Radiation Protection and Measurements (NCRP), a Congressional chartered non-profit organization that helps advise the government on radiation safety. U.S. governmental agencies use this theory to estimate the cancer deaths that will result from radiation accidents.

A second group agrees that the linear, no-threshold theory is valid at high radiation exposures; however, its members believe very strongly that radiation doses in the range of a few times our yearly background level are either harmless or are actually beneficial. To support their views, they cite extensive epidemiological studies such as the following:

- In some areas of the world, the natural background radiation level is much higher than in the United States. A survey of 70,000 people in southwest India over a 35-year period found an average background dose over three times higher than ours; some persons received 10 times our average. The study did not find any health effects resulting from the high background.

- A study was carried out in China between 1970 and 1986 in which 74,000 people in one county received 2.6 times as much background radiation as 77,000 people in an adjacent county. For the age group 10-79 years, the cancer mortality rate was about 17% *lower* in the high background group than in the other — that is, 17% fewer people had died from cancer by a given age.

- A study by researchers at the Argonne National Laboratory in 1973 found a 15% lower cancer death rate in seven western states than in the rest of the continental United States; however, the background levels in those seven states are almost twice as high as in the other 41.

- Studies of homes in China, Japan, and the United States having radon levels up to five times above average indicate that the people in those homes have *lower* cancer death rates.

- Deaths from various causes for two groups of shipyard workers employed around nuclear submarines were compared in a Department of Energy study released in 1991. 28,000 employees received radiation exposure from nuclear materials; 33,000 workers of similar age with similar jobs received no exposure. The study looked at deaths between 1960 and 1982. There was no meaningful difference in the amount of cancer the two groups had. However, the nuclear workers were much healthier than the non-nuclear workers — they had a 24% *lower* death rate from all causes than the workers who received no radiation.

Members of group two interpret these studies and many others as meaning that low level radiation is not harmful. Many members of the group go further and interpret the results to mean that radiation is actually beneficial; this beneficial effect is called radiation hormesis. Group one members, however, argue that these and other tests are not meaningful because there were not adequate controls on the studies — for example, that the Chinese group living in the county with high radiation levels also had a higher standard of living and better health care. The group one members maintain that factors such as diet and health care rather than radiation accounted for the lower mortality rate.

As indicated, there is a very sharp split in viewpoint between the two groups. Group two is supported by the Health Physics Society, most of whose 7,000 members are professionals in the field of protecting people against radiation hazards. Group two is also supported by the United Nations Scientific Committee on the Effects of Atomic Radiation; that Committee issued a report in 1994 which officially declared its belief in radiation hormesis. In the words of a former chair of the Committee, the report "dispels the common notion that even the smallest dose of radiation is harmful."

This is not an easily settled argument — whether very low doses of radiation are harmful or whether they are harmless and possibly beneficial. I have friends who are highly respected scientists on both sides of it. *One point which seems clear, however, is that any effect will be very small at worst.*

Why Do People Have an Excessive Fear of Radiation?

Contrary to the above evidence, however, many people have an excessive fear of radiation. There are many reasons for this, including the following.

1. The mass media (TV, newspapers, etc.) distort the significance of radiation. A harmless radiation release from a nuclear plant will frequently receive much more publicity than an accident involving several deaths in another industry. The Christian Science Monitor (which I otherwise respect) recently referred to nuclear power as "the most dangerous technology ever devised"; yet, less than 40 people worldwide have died from nuclear power radiation. Many people equate publicity to importance. In another example, the media frequently refer to "deadly" radiation — although no deaths occur, but do not refer to "deadly" electricity — although many people die yearly from accidental electrocution.

2. When an accident occurs, grossly unrealistic casualty figures can be reported by the media; equal publicity is seldom given to corrected figures. The Chernobyl disaster in Ukraine, the worst accident in the history of nuclear power, occurred in 1986. The Christian Science Monitor stated in April 1996 that "The casualty figures since (the accident) have mounted into the thousands." However, the International Atomic Energy Agency sampled several thousand people from the 825,000 living near the site; the general conclusion of the study, reported in 1991, was that there were *no health effects attributable to radiation*. This report received little attention in the media; most people are unaware of it. However, many of them will remember "thousands of casualties."

There *have* been health effects detected since 1991, however; a large increase in the incidence of thyroid cancer in children living near Chernobyl has been observed, and three have died. The death of three children, and several more to come (even though thyroid cancer is generally curable), is tragic; but it is a far cry from "thousands." Some of the radioactive iodine released from the damaged reactor fell on plants, cows ate the plants, and children drank their milk. The radioactive iodine in the milk concentrated in the thyroid glands, where it caused cancers. Authorities could have prevented most of these cancer cases simply by prohibiting the consumption of the milk for about three months. They could also have prevented the cancer cases if they had been able to give potassium iodide tablets to the children; the iodine in the tablets would have saturated the thyroid glands and prevented the absorption of radioactive iodine in the glands. France has recently announced a policy to have these tablets available near reactor sites as a general precautionary measure.

Thirty-one plant personnel also died within the first three months of the Chernobyl accident, 28 from acute radiation injury and three from explosion and burns.

There were also health effects due to the *fear* of radiation. It has been estimated that there was an increase of tens of thousands of abortions in Europe following the Chernobyl accident.

The absence of health effects attributable to radiation is also documented in a report by the Nuclear Energy Agency of the Organisation for Economic Co-Operation and Development dated November 1995. The members of the Organisation are 18 countries of northern and central Europe, Australia, Canada, Japan, Mexico, New Zealand, Turkey, and the United States. The report recognizes the immediate deaths and the thyroid cancers in children and possibly in adults. It then goes on to say:

"On the other hand, the scientific and medical observation of the population has not revealed any increase in other cancers, as well as in leukemia, congenital abnormalities, adverse pregnancy outcomes or any other radiation induced disease that could be attributed to the Chernobyl accident. This observation applies to the *whole general population, both within and outside* the former Soviet Union." (Emphasis added by this author.)

3. Another reason for the public's excessive fear of radiation stems from the way in which it learned about the Three Mile Island and Chernobyl accidents. Both events received tremendous amounts of publicity, and both were reported as alarming, disastrous events (which the Chernobyl event was). This publicity is not easily forgotten.

   Coupled with that is the fact that people tend to be much more fearful of a single event in which a large number of people are killed than many small events in which the same total number dies. As discussed previously, the effects of radiation from the Three Mile Island accident on the public were negligible. However, the radiation effects of Chernobyl may be large because of the number of people involved. It is estimated that people received radiation doses as follows:

   - 200,000 cleanup workers received an average dose of 33 times our normal yearly background. 600,000 additional cleanup workers received smaller doses;

   - approximately 125,000 people were evacuated; they received an average dose of five times our normal yearly background;

   - 270,000 people live in strict control zones; they received an average of 17 times our normal yearly background during the 1986-1989 period; and

- millions of people farther from the plant site received very small quantities of radiation.

Using the linear, no-threshold theory, one would calculate an ultimate cancer death toll of about 24,000 people worldwide over the next 50 years. A large number like this will readily alarm almost anyone. This is so even though the same population will have about 600 million spontaneous or "natural" cancer deaths in the same period of time.

However, the number bears further examination. First, the worst exposed group of people received average doses of 33 times normal yearly background; as discussed previously, we have no experiments to show that anyone receiving that small amount of radiation will die — thus, no experiments to show that any of the 24,000 will die. It is quite possible that low doses of radiation will prove to be harmless or beneficial. Second, about 10% of the 24,000 deaths are calculated to come from among the cleanup workers, the evacuees, and the people living in the strict control zones; the other 90% come from among the millions of people living further from the plant. If the 90% is assumed to come from among the 250 million or so people in northern Europe including the European portion of the former USSR, the chance that any one individual would die of cancer from Chernobyl radiation is about one in 10,000. Even if this crude estimate is off by a factor of 10, the chance is still only one in 1,000 of dying 20 or 30 years later from cancer. This could be of some concern, but hardly of abnormal alarm.

Further support for the argument against undue alarm comes from the fact that, beyond the strict control zones where the 270,000 people live, the highest radiation levels anywhere in the former Soviet Union and elsewhere around the world were in nearby Bulgaria. The average Bulgarian received about 32% more radiation than normal in the first year after the accident. However, if the background level in Bulgaria is equal to the world average, then the residents of

Colorado receive *over twice* as much radiation *every year of their lives* as the Bulgarians did that year. Radon, radiation from uranium, and cosmic ray levels are higher in Colorado than in most places around the world.

The numbers discussed in this item No. 3 generally refer to average doses; individual doses can vary considerably. For example, it is calculated that some of the evacuated people received as much as 135 times normal background; this compares with five times for the average person. There is obviously a higher probability for cancer death among the individuals with extreme exposures.

Summary

In summary, we must treat radiation with respect. At the high levels associated with atomic weapons, it can be quite harmful. At low levels associated with nuclear power, its effects are uncertain but certainly small and possibly beneficial. Low levels are not a cause for excessive or abnormal fears.

Comparison with Alternate Electricity Sources

Radiation is not involved in making electricity from fossil fuels. However, people living near coal plants typically receive 100 times as much radiation as those living near a nuclear plant. This is because coal has uranium, thorium, and other radioactive materials mixed in with it. When the coal is burned, the radioactive materials go out the smokestack; a relatively harmless amount of radiation is spread downwind from the stack.

Harmful Results of This Fear of Radiation

The exaggerated fear that the public has of radiation is harmful in many ways. It has led to greatly increased costs for nuclear electricity. It led to thousands of unneeded abortions following the Chernobyl accident. It holds back the acceptance of food sterilized by radiation; experts believe this sterilization could prevent thousands of deaths each year in the United States caused by food poisoning.

Chapter 6

NUCLEAR POWER PLANT SAFETY

As stated previously, the radiation released from a nuclear plant is small. However, there are large amounts of radioactive materials within the reactor. This necessitates that engineers design the plant carefully for worker and public safety. Let us examine that topic in this chapter.

As noted, the fragments resulting from fissioning the uranium and plutonium nuclei are radioactive; some emit radiation instantly, while others release their radiation over a period of minutes or years. During operation and for a considerable period after reactor shutdown, this radioactivity level is very intense. The uranium and plutonium in a reactor are located at the center in a volume about 12 feet high and 15 feet or larger in diameter; this is called the "core" of the reactor. An individual standing at the edge of the core during reactor operation would receive a lethal dose in a fraction of a second. (It is impossible to stand there.) If a large fraction of the material could somehow escape from the reactor, it could be harmful to the public as happened at Chernobyl.

The structural materials inside the reactor also become radioactive when they absorb neutrons; however, they are a much smaller source of radiation than the fuel and will be ignored in this book.

Plant workers are protected against the radioactivity in the core by massive shields made of materials such as lead, iron, and concrete. These materials surround the core and absorb most of the radiation. These are shown in Fig. 5.

A scenario by which radioactive material might escape from the reactor is as follows. The fuel rods that make up the core consist of the uranium, plutonium, and fission fragments in solid form; each rod is clad or coated with an alloy of the metal zirconium. Very little radioactive material can escape as long as the rods are in this form. However, during reactor operation, heat is generated in each rod, and it is continuously removed by cooling water. If the water flow is interrupted, the temperature of the rods will rise; if the reactor isn't shut down, the rods could melt. The molten fuel could then melt its way through piping and walls and might reach the outside of the building that surrounds the reactor. Radioactive material could be blown off the site by the wind.

Another possibility is for the heat generation rate in the rods to increase beyond the ability of the water to remove the heat. Again, melting could occur.

Fuel rod melting could also take place even if the reactor is shut down. This is because heat is generated by the radiation coming from the fission fragments; as noted, some of this radiation is not emitted until days or years after fission takes place. Melting would take place more slowly with the reactor shutdown, but it would still be a possibility, especially in the first few hours after shutdown.

Consequently, the focus of nuclear power plant safety is simply: (a) to keep the reactor running at a steady level while in operation, and (b) to keep adequate water flowing over the fuel rods so they stay cool and solid, both during operation and after shutdown. Attention is also given to preventing the release of radioactive material outside the plant boundary even if fuel melting should somehow occur.

The approach taken to assure plant safety is called *defense-in-depth:* several consecutive safety features are provided for important functions rather than just single ones. For example, to protect against loss of water flow and subsequent fuel melting,

- High-quality water pumps are used.

- Backup pumps and several supplies of water are installed to provide cooling in case the regular pumps or the water supply fail for some reason.

- Because the pumps are driven by electric motors, several sources of electricity are provided. One source is the plant itself. If the plant shuts down, electricity can be obtained from at least two separate sources outside the plant. If a hurricane destroys the power lines from both off-site sources, electricity is obtained from an emergency, diesel-motor-operated electrical generator located on the plant site. This generator is always housed in a bunker built to withstand storms, earthquakes, floods, fires, and so on. If that generator fails, there is a second and sometimes a third that can be called upon.

- In some newer plants, cooling water will be stored in tanks where it can flow by gravity (in the event of total pump failure) and provide cooling for several hours while the pumps are being repaired.

The entire plant is designed with this defense-in-depth concept in mind; there is backup instrumentation; there are multiple shutdown systems; there are multiple fire barriers; and so on. Duplicate safety systems are designed so that no common component could cause both to fail.

However, engineers recognized that failures can occur even in the best-designed plants; the possibility of rod melting is not ignored. Several physical barriers are provided to prevent the spread of radioactive material beyond the plant in the event of melting. These include the cladding around the fuel, which is made of a high-melting-point material, and a several-inch-thick steel vessel in which the core is located. Each reactor is further enclosed in a building designed to contain any radioactive material that might escape from the reactor. These buildings or domes are airtight and have several-feet-thick walls made of steel-reinforced concrete; they are designed to protect the reactor against tornadoes with 300 mile-per-hour winds, earthquakes, direct hits by large aircraft, and so on. These domes are familiar structures at plant sites. These barriers for a representative pressurized-water-reactor plant are shown in Fig. 13.

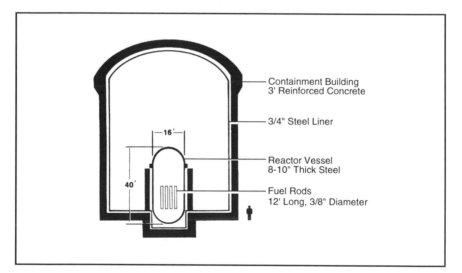

Figure 13:   Pressurized-Water-Reactor Containment Barriers
(Credit: Westinghouse Electric Company)

Plant operation is equally important to design. Quality personnel are employed, and they receive extensive and continuing training. The reactor operators must each pass repeated, periodic exams monitored by the U.S. Nuclear Regulatory Commission and be licensed continuously by that agency.

The effectiveness of our nuclear power plant safety efforts can be shown by citing our actual experience. In the 36 years of commercial nuclear power plant operation in the United States involving about 120 reactors, there has been only one major accident — at the Three Mile Island plant in Pennsylvania in 1979. The reactor was destroyed. However, as stated earlier, the amount of radiation released from the plant was so small that no plant worker or member of the public was harmed by it.

Both industry and the federal government deserve credit for this superb record. Industrial designers have incorporated many safety features in their plants. Utility organizations have established highly competent operating groups and have set up a strong organization (the Institute of Nuclear Power Operations) to continuously improve their operation. The Congress established the Nuclear Regulatory Commission to set safety standards for design, construction, training, and operation, and to oversee the entire industry.

The majority of the plants that have been built worldwide have been based on American designs and safety standards. There have been no accidents in any of these plants where members of the public have been harmed. Many other non-U.S. designs have been equally safe.

The only major accident in nuclear power's 36-year history, besides that at Three Mile Island, occurred at Chernobyl. As stated previously, the reactor failed catastrophically; much of the radioactivity in the rods was released outside the plant as airborne dust — its most dangerous form. A low level of radiation was spread over parts of the former Soviet Union and several European countries; millions of people were exposed to the radiation. It is hard to imagine a worse nuclear plant accident.

That reactor and similar ones built in the former Soviet Union are of a unique design; they are much inferior to those used in most power plants throughout the world. For example, if an American-type reactor loses its cooling water due to an accident, it shuts down; a Chernobyl-type reactor will "speed up." If an American pressurized water reactor heats up and the cooling water boils, it tends to shut down; in the Chernobyl type, the reactor speeds up when the water begins to boil. Chernobyl-type reactors are unstable, and an extremely dangerous safety problem can exist when one of them is at a low power level. The Chernobyl reactor was at such a level when the accident occurred. American industry would not design such a reactor. The Nuclear Regulatory Commission would not allow it to be built in the United States.

The plant manager, who had no nuclear training, was also a major cause of the Chernobyl accident. In his desire to run an experiment on the turbine and generator, he totally ignored the advice of the reactor staff. He further insisted the reactor be run at the low level where it was unstable and unsafe. Finally, he ordered that several control devices be manually disconnected; these devices automatically shut the reactor down when unsafe situations arise. The results were predictable and predicted; it was not an accident. Rather, it was a predictable event. The reactor was destroyed. I believe American industry would not operate a reactor in such a foolish, irresponsible fashion; the Nuclear Regulatory Commission would not allow it.

The Chernobyl plant was not enclosed in a containment building like U.S. plants are; however, this probably did not matter much. There was a steam explosion when the accident occurred, and most containment buildings would not withstand the force of that explosion.

Several Chernobyl-type reactors still operate in and around Russia; Hopefully, recent modifications have made these reactors adequately safe, although they will never reach American safety levels. It is believed that future Russian plants will be built with containment domes, and they will be designed and operated in a fashion much closer to American safety standards.

Sabotage of Nuclear Plants

You may wonder if it would be easy to sabotage a power plant and cause radioactive material to be released. The answer is "No", although full details are not known. This is because industry and the Nuclear Regulatory Commission keep the results of sabotage studies confidential; they do not want to provide information that would help a saboteur damage a plant.

Several steps are taken to protect against sabotage. For example,

- All plants have well-armed and highly trained security forces that are routinely drilled and tested.

- Access to important areas of nuclear plants is controlled by security officers who search entering vehicles and people. Individuals entering a plant must pass through metal and explosive detectors.

- The Nuclear Regulatory Commission requires that all plants have massive vehicle barriers to protect against truck bombs. The barriers also keep intruders from entering the plant.

- All new employees must pass a variety of tests and checks. These include drug and alcohol screening tests and psychological evaluation. They also include a check of employment records, criminal records through the FBI, and credit histories. Every plant has a formal program under which the behavior of all employees is monitored continually; the aim is to detect any unusual or erratic behavior.

- The Nuclear Regulatory Commission continuously consults the FBI and other intelligence agencies to evaluate potential threats.

- Plant design automatically helps protect against sabotage. The defense-in-depth design approach provides backup systems that require multiple failures before damage occurs. The reinforced-concrete containment domes protect against outside forces such as airplane crashes. The domes and the massive shields around the core provide considerable protection against external explosions — for example, bombs dropped from small airplanes or artillery shells fired by saboteur groups.

Thus, it appears that it would be very difficult for a saboteur group to harm the public by damaging a reactor.

Summary

In summary, nuclear power is safe; no member of the public has ever been killed from the operation of American-type plants. Chernobyl-type plants can not be built or operated in the United States; all existing plants will likely be phased out within a few years.

It appears that no deaths will result worldwide from nuclear power's first 36 years of history — except at Chernobyl. This is a truly phenomenal safety record for a new technology.

Comparison with Alternate Electricity Sources

A document issued by the Natural Resources Defense Council (NRDC) in May 1996 provides a comparison. The NRDC analyzed data obtained primarily from two studies: one was a Harvard School of Public Health study reported by Dr. Douglas W. Dockery of Harvard University and seven others; the second was an American Cancer Society-Harvard Medical School study reported by Professor C. Arden Pope III of Brigham Young University and six others. Both studies dealt with the effect on our health from tiny particles of matter in the air we breathe. Burning fossil fuels is the largest single source of these small particles; this includes burning coal, natural gas, oil, diesel fuel, gasoline, and wood. Coal-fired power plants are the worst offenders by far. The NRDC analysis estimates that approximately 64,000 people may die prematurely each year due to these particles; the particles cause heart and lung disease. This number is for 239 metropolitan areas in the United States; the number for the entire country would likely be around 100,000 deaths per year. The lives are shortened by an average of one to two years in the most polluted areas. One-third of these deaths are estimated to result from discharges from electricity generating power plants.

These particles are very small; some have diameters of 10 *microns,* which means that 2,500 side-by-side would be shorter than an inch. Seven side-by-side would be about as wide as a human hair.

The NRDC believes tens of thousands of premature deaths could be prevented yearly by reducing particulate emissions. It recommends switching from coal to natural gas for generating electricity; natural gas plants emit only a fraction of the particles that coal plants emit.

Nuclear plants would reduce those premature deaths even more; nuclear plants emit no such particles.

Nuclear power probably saves thousands of lives each year. Twenty percent of our electricity comes from nuclear energy; coal provides only 50% compared with the likely 70% if there weren't nuclear power. Nuclear power has thus held down the amount of pollution from coal and saved lives accordingly. The data indicate that replacing all our coal plants with nuclear plants would save tens of thousands of lives yearly.

The U.S. Environmental Protection Agency (EPA) currently regulates the emission of particles above 10 microns in diameter. Proposed new standards will place limits on particles above 2.5 microns — four times smaller; current and proposed standards also limit the number of particles in the air. The NRDC estimates that these new standards would prevent between 5,000 and 38,000 of the 64,000 deaths per year discussed previously; the actual number would depend on how strict the new standards are. If the nation adopts very strict standards, then the 38,000 number would apply. However, even the less-strict standards will likely cost billions of dollars; all will be resisted fiercely.

Others have estimated the effects of air pollution in the United States. Bernard L. Cohen, Professor of Physics and Radiation Health at the University of Pittsburgh, estimated in 1990 that fossil-fuel burning in electric power plants may cause 30,000 deaths per year by air pollution. The late John Lenihan, past Professor of Clinical Physics at Glasgow University and Regents' Professor of Chemistry at the University of California, has estimated that air pollution may cause up to 40,000 cancer deaths per year. The EPA has made crude estimates of 70,000 deaths per year associated with particle pollution.

It is interesting to note that wood, a renewable energy resource, presents considerable health hazards. Residential wood burning releases more of some kinds of particles to the atmosphere than do coal-burning power plants. Aspen, Colorado and Klamath Falls, Oregon do not meet current EPA clean-air standards because of wood smoke.

There are other risks related to power production. For example,

- 15,000 people died when the Gujarati hydroelectric dam in India failed in 1979. The Vaiont dam in Italy failed in 1963, killing 2,000 people.

- Close to 90,000 miners have been killed in coal mine accidents in the United States in this century; the toll continues at 50 or more miners killed each year.

- 1,440 people were killed in 24 natural gas accidents (fire, explosion) between 1969 and 1986 according to a recent study.

- 2,070 people were killed in 57 oil accidents (refinery fires, transportation) in the same period according to the study.

Chapter 7

HIGH-LEVEL WASTES

As with any system, nuclear plants have wastes that must be discarded; some of this waste involves radioactivity. In some materials, the level of radioactivity is trivial; the materials are harmless and can be discharged to rivers or to the atmosphere.

Other material is too radioactive to discharge in that manner but is only modestly hazardous. It is buried underground along with similar wastes from hospitals and radiochemical laboratories. Because this material does not represent a health hazard when properly disposed of, it will not be discussed further.

A third class of wastes, the used or "spent" fuel rods from a nuclear reactor, presents a greater challenge; the rods become intensely radioactive during their three- to five-year residence in the reactor. They must be disposed of carefully after they are discharged. Because of its radioactivity, the material in the spent fuel is termed *high-level waste* or *HLW*.

HLW Disposal Methods

When spent fuel rods are discharged from a reactor, they are stored in water-filled concrete pools similar to deep swimming pools. This is a satisfactory storage method for several years. However, it is unsuitable for long-term storage.

Many methods for long-term disposal of HLW have been suggested. They include burying it 30 feet or so below the ocean floor where it would be isolated from humankind; burying it near the South Pole where (since it emits heat) it would melt its way down through several thousand feet of ice; and shooting it into outer space on rockets. However, the best approach appears to be to bury it beneath the surface of the earth — in the ground. Burial could be in volcanic, salt, granite, or other layers of material below the earth's surface. Most nations are pursuing this approach. Sweden is considering granite layers; Germany is considering salt layers; and the United States is investigating volcanic layers at Yucca Mountain in Nevada. Congress passed a law in 1982 requiring the U.S. Department of Energy (DOE) to handle the burial of HLW from commercial nuclear plants. Burial was to begin in 1998, but the program is several years behind schedule. The DOE will be required to provide temporary storage before burial begins.

Composition of HLW

Spent fuel material inside the cladding is composed of about 95% uranium, 1% plutonium, 3.5% fission fragments, and less than 0.1% of other elements that are heavier than uranium. There is also oxygen in the material, but it is unimportant in our discussion.

As you will recall, uranium occurs in nature, and fission fragments are formed when nuclei split during fissioning. You will also recall that plutonium is made when U-238 nuclei capture neutrons; plutonium, in turn, will usually fission when it captures a neutron. However, sometimes it will simply absorb the neutron and form the next-heavier element, americium. The latter, in turn, can absorb a neutron and form the still-heavier element, curium, and so on. There are about a dozen of these man-made, heavier-than-uranium elements; plutonium is obviously one of them. They are called "TRUs," which stands for transuranic elements. These elements are similar to the 92 in the periodic table; all have nuclei composed of protons and neutrons that are surrounded by electrons. However, none exists in nature. Most can be made to fission like U-235.

Hazards of HLW

The composition of the spent fuel (uranium, fission fragments, plutonium, and other transuranic elements) determines the difficulty of HLW disposal. Uranium is only a relatively small health hazard and can be buried with little concern.

The fission fragments are intensely radioactive when formed and when the rods are discharged from the reactor. However, there is something special about all radioactive substances — *they all lose their radioactivity as time passes.* This makes it easier to dispose of the fragments. Let us examine this special feature.

Natural Decrease of Radioactivity Levels

Assume that a radioactive material is giving off 1,000 particles of radiation each minute. Now, if we watch it, we will find that the level will fall in half (to 500 particles per minute) eventually; we say the radiation level *decays*. Maybe it took four years for the level to fall in half. Let us call that four years the *half-life* of the material. If we keep watching, we will find that the level will fall in half again or to 250 particles per minute in another four years. The process will continue. In four more years, the level will fall to 125 particles per minute. And so on. In 10 half-lives or 40 years, the level will decrease by a factor of 2 x 2 x 2 (10 times) = 1,024 or to about one particle per minute.

Note: When many materials decay by radiation emission, they become new elements. A new element may be stable (that is, not radioactive) and will exist forever, or it may be radioactive. If it is radioactive, it, too, will decay. This process will continue until a stable element is reached. This explains why transuranic elements do not exist in nature. Many existed when the earth was formed, but they are all radioactive. Consequently, all have decayed into new, stable elements — they have become elements among the 91 stable elements in the periodic table. The main isotope of plutonium has a half-life of 24,600 years, but the earth is a few billion years old. Therefore, for all practical purposes, natural plutonium does not exist. (A little exists, but the amount is becoming infinitesimally small as time passes.)

Different fission fragments have different half-lives. Some have half-lives of fractions of seconds; others have half-lives of days or thousands of years. However, all the fission fragments together have an effective half-life of about 30 years; their activity will decrease by a factor of about 1,000 in 300 years. Their total activity would not be a serious health hazard at that time. Further, it would not be difficult to develop practical burial methods to isolate these materials for that period of time. This contrasts with many chemical wastes; for example, the mercury in our rivers will remain toxic forever. Two fission fragments, technetium-99 and iodine-129, have half-lives over 200,000 and 15,000,000 years, respectively, but they are present only in very small quantities.

Hazards of Plutonium and Other Transuranic Elements

The third component of HLW, plutonium, presents a more serious waste-disposal problem. As stated previously, the most plentiful isotope of plutonium, Pu-239, has a half-life of 24,600 years, and it is a health hazard under some conditions.

Plutonium decays by emitting alpha particles; they travel less than 1.5 inches in air, and they will not penetrate the skin. Therefore, plutonium is not a hazard as long as it remains outside the body.

If plutonium gets inside the body, its alpha particles lose all their energy in a very short distance; they can be very damaging to sensitive tissues nearby. In significant quantities, it can cause injury if it remains within the body. Injury can include cancer many years later.

Plutonium can enter the body through the mouth (food, water), nose (breathing airborne particles), or cuts and wounds. Studies indicate that small quantities taken in through the *mouth* pass through the digestive tract with very little being absorbed. Some Japanese groups state publicly that solutions of plutonium can be drunk without harm. Professor Cohen of the University of Pittsburgh has offered to eat about a gram of plutonium to demonstrate that eating it is no more dangerous than eating the same quantity of caffeine. Plutonium and radium act similarly inside the body; plutonium is about one-sixth as poisonous as radium on a weight basis. It was noted earlier that only about 2% of the radium-dial painters who ingested large quantities of radium died later from cancer.

Studies on dogs indicate that *breathing* airborne plutonium can be serious; very small amounts administered to beagles consistently caused lung cancer. However, these studies do not seem to apply to humans.

Humans appear to be less harmed by plutonium than animals, although human data are limited. Many workers in the Manhattan ("Atomic Bomb") Project in the 1940s got plutonium into their nostrils; however, they apparently developed no more lung cancer than the rest of the population. Twenty-six men inhaled plutonium (or absorbed it through cuts) at the Los Alamos National Laboratory in the mid-1940s. At their last examination in the early 1990s, seven of the 26 had died; this is less than the 13 deaths that would be expected normally. Plutonium still shows up in the urine of the 19 survivors and always will.   Eighteen seriously ill hospital patients were injected with small doses of plutonium in the 1945-1947 period; five of the subjects were still alive almost 30 years later, and no ill effects of the plutonium were observed.   Thus, plutonium in large quantities will surely cause some cancer deaths just as radium does; however, *there has never been a known case of death resulting from plutonium.*

Because of plutonium's health effects, HLW containing large amounts of plutonium must be isolated from humans.   One effective way to do this is to bury the waste deep underground.

Many of the other transuranic elements also have long half-lives and emit potentially harmful radiation.   In particular, neptunium-237 has a half-life of 2,140,000 years and americium-243 has a half-life of 7,370 years; both are alpha emitters like plutonium.   Although small, the quantities of such transuranic elements are large enough to warrant disposal methods similar to those for plutonium.

Burial of HLW

As stated earlier, current U.S. plans are to bury the HLW underground, and Congress assigned the Department of Energy the responsibility for (a) accepting the spent fuel from nuclear power plants by January 1998, (b) for building an underground disposal facility, and (c) for accomplishing the burial. The DOE is currently studying Yucca Mountain to show that that site is suitable for HLW burial, and present planning calls for the spent fuel elements to be loaded into massive concrete and steel casks that will be buried about 1,000 feet below the surface of the Mountain.   A 1990 report of the National Research Council (which is administered by the most prestigious scientific and engineering bodies in the United States) states:

"There is a worldwide scientific consensus that deep geological disposal, the approach being followed in the United States, is the best option for disposing of high-level radioactive waste (HLW). *There is no scientific or technical reason to think that a satisfactory geological repository cannot be built.*" (Emphasis added by this author.)

This DOE burial program is behind schedule for many reasons. People in Nevada do not want the wastes buried there — the "not in my backyard" syndrome. It is also easy for any group opposed to burial or to the use of nuclear power in general to delay the program; delay is readily accomplished by challenges through our court system. As this book is being written, Congress is endeavoring to pass legislation to bring the burial program back on schedule.

It is interesting that the plutonium formed in the ground in the natural reactor in Gabon has moved less than six feet from where it was formed 1.8 billion years ago; the plutonium lies loose in the ground, of course, in a tropical-rainfall region. Plutonium sticks strongly to soil with which it comes in contact.

Transportation of Spent Fuel Rods

The burial of HLW will obviously involve the shipment of spent fuel rods from power plants to the burial site. Many people ask: Are such shipments safe?

The handling of spent fuel is based on the defense-in-depth concept just as reactor safety is. As stated earlier, when spent rods are discharged from the reactor, they are first stored in water-filled concrete pools for several years. During this time they lose about 95% of their radioactivity.

Then, they are shipped in carefully designed, fabricated, and tested shipping containers. These massive containers are normally 15 to 20 feet long with foot-thick walls; they weigh 25 to 40 tons for highway shipment and up to 125 tons for rail shipment.

The containers or "casks" must be designed to meet Nuclear Regulatory Commission requirements and be licensed by that agency. They must be able to withstand all of the following events, one after another:

- the equivalent of being dropped several hundred feet onto a hard surface,

- being immersed in a 1,475 °F fire for 30 minutes, and

- being submersed under water for eight hours.

Engineers at Sandia National Laboratory have tested casks of this type extensively. In one example, a locomotive speeding at 80 miles per hour smashed broadside into a cask parked on the railroad track. The locomotive was demolished, but the cask suffered only negligible damage. No material escaped from the cask. In another example, a cask was mounted on a railway car that crashed into a concrete wall at 80 miles per hour. The cask was then surrounded by a fire that was hot enough to melt all the lead in the cask. Again, the cask received only minor overall damage. Another cask was dropped 2,000 feet onto packed ground as hard as concrete. It was traveling 235 miles per hour when it hit the ground, and it buried itself four and one-half feet deep. The only damage was to the paint on the surface.

Additional steps are taken to ensure transportation safety. For example, routes proposed for highway shipment are submitted to the Nuclear Regulatory Commission for approval. Escorts are required, and they must be trained in physical protection of the shipment; at least two armed escorts are required in heavily populated areas. The escorts must call the communication center every two hours. For highway shipments, the driver must be given special training in security procedures.

The National Conference of State Legislatures recently issued a report on transporting spent fuel. It states that over 2,500 shipments have been made, with no deaths or injuries due to any radiation-related cause. Most of the shipments were by truck, some were by rail.

Thus, our safety record in shipping spent fuel is outstanding; it should continue this way.

Waste Treatment in Other Countries

There is a significant difference between how we plan to dispose of HLW and how most other nations do. Instead of burying whole spent fuel rods, other nations intend to chemically process (dissolve) the rods and remove the uranium and plutonium. These metals will be reused in the reactor, and only the fission products and remaining transuranics will be buried. It was American policy to do this also in Presidents Eisenhower's, Kennedy's, Johnson's, and Nixon's administrations.

However, dissolving the rods under the process currently used leads to the presence of pure plutonium. Plutonium is relatively easy to handle, and it can be the major ingredient in atomic bombs. Officials in Presidents Ford's and Carter's administrations in the late 1970s became concerned about separating the uranium and plutonium; they worried that the existence of pure plutonium at several locations worldwide would lead to the theft or diversion of the material by terrorist organizations and rogue nations. Consequently, the policy was established under President Carter that the United States would not dissolve rods before burial; it was hoped that other nations would follow our lead, but that did not occur. President Reagan reversed the policy in the 1980s. By that time, however, American industry lost interest in the process, and we have no commercial facilities of this type in the United States. Therefore, American practice will be to bury the whole rods. Dissolution facilities could be built, but they are expensive and uranium is cheap at this point in time. Our nuclear industry could theoretically send our spent rods to other countries for dissolution; however, this would require complicated federal approvals. No utility organization seems interested in pursuing that possibility. In burying the HLW, we risk losing large quantities of uranium and plutonium — a valuable energy source. Many people strongly oppose burial for this reason — believing that it is too wasteful. Recovery of buried fuel would be difficult.

Many nations, including Belgium, England, France, India, Japan, and Russia, currently do or intend to reprocess spent fuel. They will bury only fission products and transuranic elements other than plutonium. This approach will allow them to utilize the energy in the uranium and plutonium from the spent fuel. It will also mean that there will be much smaller quantities of waste to bury; the volume of the fission fragments and transuranic elements (other than plutonium) is only a small fraction of that of the unprocessed spent fuel rods.

Future Developments

Work is underway at Argonne National Laboratory that could lead to simplification of HLW disposal; a new method of processing spent fuel is being developed. In this process, the uranium is separated and reused; the plutonium, the other transuranic elements, and the fission fragments would all be mixed with glass and buried underground. The volume of material being buried would be reduced significantly compared with burying complete fuel rods.

Further, with one modification to the process, the fission fragments could be separated from the plutonium and other transuranics. The transuranics could then be cycled through a "fast reactor" (described in the next chapter), fissioned, and destroyed — only fission fragments would remain. All fission fragments would be buried subsequently. Because the effective fission-fragment half-life is only about 30 years (as noted earlier), the waste would be no more hazardous to our health in about 300-500 years than ordinary uranium ore as found around the world.

This reduction in the effective half-life of the waste would simplify the disposal problem technically. In addition, it would surely ease present public concerns about burial. Although the current burial program appears fully acceptable technically, a new approach may conceivably be useful if public fears persist.

At present, we do not have fast reactors in which to destroy the transuranic element waste as described. However, as will be discussed, such reactors could readily be built.

Summary

In summary, the problems in disposing of HLW are political rather than technical. Underground burial represents a satisfactory solution for disposal.

Comparison with Other Energy Sources

The fuel requirements for nuclear plants are significantly smaller than for plants using other fuels or sources of energy. This is shown in the following table for an example city.

The annual requirement of 33 tons of uranium fuel can be shipped in a few railroad boxcars. Shipping 2,300,000 tons of coal requires about 214 trains of coal, each train being about 105 cars in length. This is equivalent to a single train 250 miles long.

The quantity of waste discharged from a nuclear plant is also significantly smaller than from a coal plant. Dr. Hans Blix, Director General of the International Atomic Energy Agency, compared the wastes from a coal plant having optimal pollution

Yearly Fuel Requirements for a Power Plant Generating
Enough Electricity for a City of 560,000 People.
(Credit: Department of Energy)

Fuel	Requirements
Uranium	33 tons
Coal	2,300,000 tons
Oil	10,000,000 barrels
Natural gas	64,000,000,000 cubic feet
Solar cells	39 square miles
Garbage	7,000,000 tons
Wood	3,000,000 cords

abatement equipment with the wastes from a nuclear plant. The
plants were approximately the size of the plants in the table
above, and Blix's figures are given in the table below:

Yearly Wastes Discharged from Power Plants
Generating 1,000 Megawatts of Electricity

Wastes	Coal Plant	Nuclear Plant
Sulfur Dioxide, SO_2	1,000 tons	0
Nitrogen Oxides, NO_X	5,000 tons	0
Particulates	1,400 tons	0
Carbon Dioxide, CO_2	7,000,000 tons	0
Ashes	Up to 1,000,000 tons	—
Spent Fuel	—	38 tons

Notes:

1. The SO_2, NO_x, Particulates, and CO_2 are discharged to the atmosphere.

2. The quantity of ashes will depend on the quality of the coal. It is reported that as much as 50% to 75% of the ashes are put to use; one use is in concrete for highways.

3. A million tons of ash can contain hundreds of tons of toxic heavy metals (arsenic, cadmium, lead, mercury).

The volume of waste from the nuclear plant is also small. If the spent fuel is chemically reprocessed, the yearly volume of highly-radioactive waste will be about three cubic yards; this is about one-fifth the size of an automobile. The entire nuclear chain supporting the plant, from mining through operation, will generate an additional 800 cubic yards of lower-level waste per year — a volume smaller than 50 automobiles. If the spent fuel is not reprocessed, the volume of highly-radioactive waste will increase to about 25 cubic yards — the size of about two automobiles. The spent-fuel waste, whether reprocessed or not, will be encased in shielding before underground burial; the volume of waste and shielding together will be 10 or 20 times the volume of the waste alone.

Dr. Blix states his viewpoint about nuclear wastes as follows:

"The issue of safe disposal of nuclear waste that remains radioactive for tens of thousands of years needs to be put into perspective. The argument has been made that it is irresponsible to leave any long-lived radioactive waste behind us. That argument, in my view, would apply with much greater strength to the toxic chemical residues–such as arsenic, mercury, lead, and cadmium– that result from the burning of fossil fuels. Their impact on health and safety is often more immediately drastic, and they do not have half-lives. *They remain toxic forever.*

The reality is that we *must* leave some waste behind us, if we want to maintain or create high living standards. The questions rather are: How do we *minimize* these wastes, and how do we make sure that they do not cause harm? The main problem with the wastes of fossil fuels is that they are so voluminous that they cannot be taken care of. Their final disposal sites are the surface of the earth and the atmosphere we breathe! On the other hand, nuclear waste, because of its *limited* volume, can be put back in the crust of the Earth from where the uranium originally came. In my view, we should talk not only about alternative energies, but also about "alternative wastes." The limited volume of nuclear wastes, I submit, is one of the greatest assets of nuclear power."

Chapter 8

DIVERSION OF NUCLEAR MATERIALS

Public acceptance of nuclear power requires satisfactory answers to questions such as: Can terrorists steal uranium or plutonium from nuclear power plants to make bombs? Could they make a bomb if they had the material? Could hostile nations make weapons from their own spent nuclear fuel? These questions are reasonable because explosive devices can be made from small amounts of either U-235 or plutonium; amounts in the range of 20 to 50 pounds are adequate, depending on the choice and purity of the material.

Fuel Rod Theft

Let us start by addressing the question of whether terrorists could steal fuel rods and make bombs. We will consider the situation in the United States initially.

The first answer is that the uranium in fuel rods can not be used to make a bomb. The U-235 content in new or fresh rods is raised or "enriched" from the natural 0.7% to about 4% or 5%. However, a mixture of U-235 and U-238 must consist of at least 20% U-235 to be explosive. Further, the U-235 content in spent fuel rods is less than in fresh ones; U-235 is consumed as the reactor is operated.

The second answer is that it would be very difficult — maybe impossible — for a terrorist group to make a bomb from the plutonium in fuel rods. There are many reasons. First, there is no plutonium in fresh rods; the terrorists would have to steal spent fuel to get plutonium. Second, spent fuel is intensely radioactive and is transported in massive casks, as described earlier. A terrorist group would have great difficulty stealing such fuel. It would have further difficulty avoiding the intensive police manhunt that would follow. It would no doubt be faced by a massive search-operation employing the most sensitive detection equipment available.

Even if it succeeded in stealing spent fuel, the terrorist group would then have to separate the plutonium from the other materials in the rods; rather-pure plutonium is required for a bomb. The group would have considerable difficulty doing so, partly because of the dangerous radiation levels involved. A person standing a few feet from a single, freshly discharged rod would receive a lethal dose of radiation in a fraction of a second. This would remain true even after the rod had been out of the reactor for a few years. Heavy shielding and complex robotic equipment would be required to protect workers during the chemical-separations process.

Further, the terrorists would need competence and thorough understanding in a wide range of technical specialties before they could make a bomb. These specialties would include implosion hydrodynamics, critical assemblies of nuclear components, chemistry, metallurgy, machining, electrical circuits, explosives, radiation protection, and others. At least several people who could work together as a team would be required; they would have to be carefully selected to ensure that all necessary skills were covered.

Costs would be high. These would include support for personnel over a period adequate for planning, preparation, and execution; surely, years would be required. A wide variety of specialized equipment and instrumentation would also be needed.

The group would encounter numerous hazards besides radiation; these would include the possibility of a premature nuclear explosion and handling conventional explosives.

There is adequate information in our libraries to tell a group how to make a bomb — to understand what must be done. However, practical problems usually arise when any complex device is made for the first time. Consequently, the group would not be assured of a successful explosion on a first attempt. Police authorities would no doubt stop a second effort.

Thus, for the previously mentioned reasons alone, I believe it is extremely unlikely that a terrorist group in the United States could make a damaging bomb from stolen fuel rods. Dr. Luis Alvarez, a scientist who worked on the first atomic bomb, said in 1987 that "making (a plutonium bomb) explode is the most difficult technical job I know."

However, there is an additional obstacle the group would face. Pu-239 is made in a reactor when U-238 nuclei capture neutrons. This isotope of plutonium is the ideal material for making plutonium bombs. If the Pu-239 remains in the reactor for a long period, undesirable impurities build up. For that reason, fuel rods remain in military reactors for only a few weeks or months. The rods are discharged before impurities can form.

Commercial nuclear power plants are operated differently. Because new fuel rods are very costly, they are left in the reactor for three or four years. During that time, many impurities build up. One of the most significant is Pu-238; this isotope emits large quantities of heat.

Plutonium bombs are made by surrounding plutonium with high explosives like dynamite. When the high explosive is detonated, the inward pressure causes the density of the plutonium to increase; this increase is enough to cause the plutonium to detonate. However, the heat generated by the Pu-238 is so intense that it would probably cause the high explosive to melt; this would happen long before the bomb could be used. The melted high explosive would not detonate, and so the plutonium would not either. It would be very difficult for a terrorist group to get around this heating problem. Competent scientists and engineers have given considerable thought to how melting could be prevented; none has found an easy answer. There is no simple solution.

In actuality, much of the Pu-238 is formed after the fuel rods are discharged from the reactor. The transuranic element, curium, is formed in the reactor as stated earlier; it then decays to Pu-238 with a 163 day half-life. Within a year after the rods are discharged, adequate Pu-238 is formed to cause melting. Spent fuel rods are stored in spent-fuel-storage pools for at least a year before leaving the reactor site. Further, it is almost impossible for a terrorist group to steal large quantities of spent fuel from a storage pool.

You may logically wonder if the Pu-238 can be removed from the other plutonium in spent fuel; the practical answer is "No." Elements such as uranium and plutonium can be separated by chemical processes. However, except in rare instances, isotopes can not be separated chemically. It is much more difficult to separate isotopes than elements. Isotope separation is beyond the capability of any "ordinary" terrorist group.

Thus, terrorists almost certainly can not use the plutonium in normal spent fuel from nuclear power plants to make a bomb.

Separation of Plutonium in Spent Fuel

As noted earlier, many countries chemically process their spent fuel and separate the plutonium. The plutonium is then used to make new fuel rods; it is substituted for part of the U-235 in the rods. The United States may do the same in the future. The question then arises: Would use of this process change the likelihood of terrorists making bombs from stolen fuel rods?

In one respect, the terrorist's job would be easier; fresh fuel rods containing plutonium would be a new target for theft. Fresh fuel rods would be much less radioactive than spent fuel; handling them would be simpler than handling spent fuel. One or two hundred shipments of fresh rods might be made yearly from fuel fabrication plants to power plants. However, our federal agencies would no doubt require stringent security measures to safeguard them. Thousands of plutonium shipments have been made in the United States in the last 50 years without apparent theft. I believe the terrorist would be no more successful than before.

Related to plutonium in fresh fuel rods is the fact that separated plutonium would exist at the reprocessing and fuel fabrication plants. Could a terrorist group make bombs from pure plutonium stolen from these sites? Again, I believe the practical answer is "No." There would be only two or three reprocessing centers and, at most, only a few fuel fabrication sites. With the stringent federal regulations we would no doubt have, the possibility of theft would be very small. Even if theft occurred, the terrorist would still face the other obstacles to developing a successful bomb.

Activities in Other Nations

Theft of spent fuel by terrorists may be more probable in other nations than in the United States; some nations may have less-stringent security measures than we do. Even so, the other obstacles to building a successful bomb would still exist.

Successful plutonium diversion would be more likely by a nation rather than by a small terrorist group. A nation having nuclear power plants would have access to its own spent fuel. It could bypass several of the obstacles discussed previously toward making a bomb. In particular, it could discharge fuel rods from its own reactors a few months after the fuel was loaded; then, there would be little Pu-238 or other impurities in the plutonium.

Consequently, there has been a concerted political effort by several leading nations (including the United States) to control the diversion of nuclear materials. This has led to actions such as the following:

- Most nations of the world have signed a Nuclear Non-Proliferation Treaty. Under the Treaty, each nation agrees to open its commercial nuclear power activities to inspection. The International Atomic Energy Agency, headquartered in Vienna, performs these inspections. It is hoped that all nations will follow the Treaty and use their commercial power plants for power production only.

- The United States has tried to limit the reprocessing of spent fuel, as discussed earlier.

- Most advanced nuclear countries have refused to sell nuclear plants to a few rogue nations such as Iraq. They have also refused to sell equipment useful for making bombs. The United States has been a leader in encouraging the advanced nations to withhold such sales.

These political efforts have been successful as a whole. Not all individual efforts have fully succeeded — for example, several nations, including India and Pakistan, have not signed the Non-Proliferation Treaty, and few nations followed our decision not to reprocess spent fuel. However, I do not know of any case (except possibly India) where a nuclear power program has significantly assisted in the development of bombs.

Further, no country is likely to make plutonium in a commercial power plant — there are simpler ways to obtain it. A power plant is large, complex, and expensive; a U.S. plant costs in the range of a billion dollars. In contrast, plutonium can be made in a simpler reactor that might cost only a fraction of that amount. Further, American-type power reactors require expensive, enriched fuel to operate; some simple reactors can use cheap, unenriched natural uranium. Thus, a nation wishing to make plutonium for terrorism purposes would normally choose to go the simple-reactor route. Still another strong reason for using a simple reactor is that the rogue nation would likely wish to build its bombs in secrecy. A simple reactor can probably be built secretly; the entire world will usually know if the country purchases and builds a commercial plant.

In fact, North Korea has followed this route in attempting to manufacture plutonium in the last few years. It has no nuclear power plants, but it has built simple reactors and a chemical-reprocessing facility. U.S. intelligence agencies learned of this effort; our government is currently negotiating with the country to get it to stop its plutonium program.

A dozen or so nations in the world are believed to have plutonium nuclear weapons. All of them built their weapons before they had power plants. The United States, for example, built atomic bombs in the 1940s; it did not have nuclear power until the 1960s. (India may have used heavy water, a component of some non-American-type reactors, to build a bomb. It argues that the bomb is for peaceful purposes such as canal building.)

Diversion-Proof Fuel

It has recently been proposed that all nuclear power plant spent fuel throughout the world be made useless for making bombs; this would be accomplished by putting Am-241, an isotope of americium, in all new fuel rods when they are made. That isotope will capture neutrons in a reactor and form Cm-242, an isotope of curium. The curium would decay to Pu-238. (This is partially the process described earlier.) The americium would be obtained from aged, existing spent fuel by chemical separation; a suitable amount would then be inserted in every new fuel rod made anywhere. By building time delays into the system, enough Pu-238 would be formed to render any spent fuel useless for bombs. Even nations with their own power plants would be unable to discharge impurity-free spent fuel after a short exposure in a reactor.

All nations would have to agree to follow such a plan for it to be successful; they would also need to agree to open their fuel fabrication plants to inspections. This would require a political treaty. Fortunately, there are few fuel fabrication plants in the world, and policing would be relatively easy. As indicated, this concept is currently in the proposal stage; time will be required to evaluate its merit.

Related Background Material

There are other serious concerns about the use of nuclear bombs in the world today. The media frequently fail to state the source of the concerns clearly. Consequently, the public often incorrectly ties them to nuclear power. For that reason, it is worthwhile to explain the concerns briefly. Note specifically, however, that *none of them is related to the use of nuclear power in the United States.*

Additional Rogue Nation Activity

Rogue nations may be able to make U-235 bombs as well as plutonium bombs secretly. Natural uranium can be enriched to the 20% U-235 level or above by several processes. None are simple, but several nations, including the United States and the former Soviet Union, have uranium bombs in their stockpiles. Uranium bombs are much easier to design and fabricate than plutonium bombs. Dr. Alvarez, who was mentioned earlier, also said in 1987 that — if a supply of highly enriched U-235 were available — "even a high school kid could make a bomb in short order."

Iraq was endeavoring to make enriched uranium until it was stopped by the 1991 Persian Gulf War. The United States led this war; one reason for doing so was to halt the Iraqi effort. Note that Iraq does not have nuclear power plants.

Surplus Uranium and Plutonium

With the end of the Cold War, the United States and Russia have agreed to reduce the amount of highly enriched uranium and plutonium that each has stockpiled; these materials were made in special military facilities for weapons purposes. There is fear that some of these materials, particularly those in Russia, could be stolen or diverted.

The United States is buying many tons of the Russian enriched uranium; this will be mixed with natural uranium to a level of 5% U-235 or less. The mixture will then be used in commercial power plants. However, this use is simply a way to consume a potential bomb material; U.S. nuclear power plants will contribute to solving an international problem. They are not dependent on the availability of such Russian uranium.

The United States and Russia have also agreed that each will remove about 37 tons of plutonium from their military stockpiles. There is disagreement between the two nations on what to do with the material. Our government is considering the following two methods to dispose of our plutonium: (a) mixing it with intensely radioactive fission products and burying the mixture deep underground, and (b) mixing it with uranium and using it in the fuel rods of commercial power plants in the manner discussed earlier. In item b, some of the plutonium would be consumed as it is fissioned in a reactor; the remainder would be buried in spent fuel rods. Note here, too, that the item b approach is simply a way to consume a potential bomb material; again, U.S. nuclear plants would contribute to solving an international problem. Our plants are not dependent on the availability of such plutonium; in fact, our government would likely have to pay the plant owners for using the plutonium.

The U.S. government is endeavoring to set an example and is encouraging the Russian government to dispose of its plutonium in a similar fashion. However, plutonium is about 10 times as expensive as gold and can be used advantageously in advanced reactors. The Russian position is that it intends to store the material and use it in such reactors at some later date. Thus, final use or disposal of their material is uncertain at present.

There are occasional articles in the newspapers about people being arrested in Poland and elsewhere for selling small quantities of uranium or plutonium on the black market. These materials apparently all come from Russia. However, they come from laboratories and institutes rather than from military stockpiles.

Surplus Weapons

Some of the U.S. and Russian surplus material exists in the form of finished, tested bombs — for example, artillery shells composed of nuclear explosives. Each country has around 10,000 bombs and there is fear that some could be stolen in Russia. This is particularly so because the Russian army is going through a difficult period. Troops are reportedly underpaid and frequently not paid; morale is supposedly very poor. Such conditions could lead to theft and diversion.

Factors Beyond Our Control

It is sometimes stated that, if the United States abandoned nuclear power, the rest of the world would also; all of the world's nuclear problems would disappear. This is obviously wishful thinking. Most of those problems are related to military applications and are independent of nuclear power. In addition, abandoning nuclear power would intensify other problems such as those arising from burning fossil fuels.

Further, nuclear power use is very important to much of the world — many nations can not or will not abandon it regardless of what we do. We can offer leadership on its safe use, and most nations will follow us — as they have done. We can also exert leadership in getting other nations to conduct their power programs openly — to allow international inspections. We led in establishing the International Atomic Energy Agency; most nations have given it authority to inspect their civilian programs. Thus, we can have a positive influence on nuclear power worldwide. However, we can not lead other nations to take actions against their own best interests — as our efforts to halt the chemical processing of spent fuel demonstrated. Fortunately, few of the world's pressing problems arise from commercial nuclear power.

Summary

In summary, we need not fear the possibility of diversion of nuclear materials from the U.S. nuclear power program. Of course, we should continue to be vigilant with our safeguards. On the international level, the United States and other nations must continue political and diplomatic efforts to control the use of nuclear materials; however, this need is independent of the use of nuclear power in the United States. There are legitimate concerns about the diversion of explosive materials and devices — particularly from Russia's military program; however, *they are unrelated to commercial nuclear power.*

Chapter 9

ADVANCED REACTORS

Today's nuclear power plants can be compared with early automobiles, early airplanes, and early TV reception. A leading airplane in the 1940s, the DC-3, carried about 50 passengers at speeds up to 200 miles per hour for 500 miles. The 747 today can carry 450 people at 550 miles per hour for nonstop distances of 5,000 miles. A typical TV set in the mid-1950s could pick up stations 75 miles away in black and white. Today's sets can bring in stations worldwide via satellite in color. Today's nuclear plants are "early" models — mostly first-generation plants. Newer models will show improvements over time, although perhaps not as dramatic as those for the other examples cited.

We can expect changes of many kinds, but we will focus our discussion on two types — evolutionary improvements in the boiling-water and pressurized-water reactors described earlier, and the development of so-called "breeder" reactors. Let us look at each.

Evolutionary Changes

American efforts have been underway for several years to design improved, second-generation plants. Participants in this program include the utility industry, which owns and operates the present plants; the manufacturers of the plants; and the Department of Energy. Foreign groups have also been involved. The efforts are aimed at designing plants that will be simpler to build, be easier to maintain, require fewer operators, be even safer, and generate electricity at lower costs than first-generation plants.

These second-generation units will also be standardized as is done in France. Instead of each utility buying a unique plant as was done earlier, there will be a choice of only a few designs from which to choose. This will also simplify operator training.

A second, evolutionary thrust is to simplify the U.S. licensing process. When a utility organization wished to construct and operate a nuclear plant in the past, it was required to

- obtain permission from the U.S. Nuclear Regulatory Commission to build a specific plant at a specific location,

- purchase and construct the plant, and

- obtain permission from the Commission to operate the plant.

The last step is controversial. A utility must borrow money to build a plant. However, there is no assurance that the plant can be operated and that the lender will get his or her money back until after the plant is built and the money is spent. In addition, groups opposed to the plant or to nuclear power in general can institute legal suits to delay the granting of permission. Such actions can be very costly to the utility; if a $2 billion plant sat idle and if the utility had to pay interest on the $2 billion at the annual rate of 12%, the delay would cost $240 million per year. The cost in four years would exceed a billion dollars. Unnecessary delays are believed to have added over a billion dollars to the cost of the Seabrook plant in New Hampshire.

Industry, the Nuclear Regulatory Commission, and Congress are currently considering ways to eliminate the controversial third step. The thrust of the effort is as follows:

- The utility would apply to the Commission to build a standardized, preapproved plant at a preapproved site.

- The Commission would grant permission prior to construction of the plant for the utility to build and operate the plant. Permission would require that the plant be constructed properly, of course.

These evolutionary thrusts are beginning to show concrete results. The General Electric Company (GE) has worked with Toshiba and Hitachi Corporations in Japan to design an Advanced Boiling Water Reactor plant. The first such plant was recently constructed for Tokyo Electric Power Company in less than 4.5 years and below the budget. In contrast, U.S. plants built in the 1980s required as much as 11 years before they could be operated, and many had huge cost overruns. A second Japanese plant is nearing completion, and additional units are planned. GE is supplying the nuclear reactors, the fuel rods, the turbines, and the generators. Taiwan has also announced plans to construct two such plants. Westinghouse Corporation is working along a similar path. South Korean organizations are utilizing an advanced pressurized water reactor developed by the ABB-Combustion Engineering Corporation (ABB-CE).

The Nuclear Regulatory Commission has completed its detailed safety reviews of the GE and ABB-CE standardized designs and given Final Design Approval for each. It is believed that Commission approval has been quite important for other nations to accept these plants. U.S. safety standards are highly respected worldwide.

Breeder Reactors

As discussed earlier, part of the energy released in a reactor comes from fissioning U-235. In addition, some neutrons released in fissioning are captured by U-238 to make plutonium, and that plutonium is also fissioned. However, there is a net decrease in fissionable material as we operate the reactor; more U-235 is used than plutonium is made. We are using up our supply of U-235 which exists in natural uranium. Apparently there is not enough natural uranium in the world to meet international needs for more than a century or so. Plenty of uranium actually exists, but the cost of recovery becomes prohibitive after we mine the easy-to-reach material.

However, there is a different type of reactor in which less U-235 is consumed than plutonium is made — there is a net *increase* in fissionable material. This is the so-called "breeder" reactor — it produces more fissionable material than it consumes. Such reactors would greatly extend our supply of nuclear fuel; close to 100% of the uranium would be made fissionable. The use of breeder reactors could meet the world's electricity needs for many centuries.

Breeder reactors usually use the metal sodium (element 11) as the coolant. Sodium melts at about 208 °F (vs. 32 °F for water, of course); because the reactor is much hotter than that, the sodium is melted into a liquid. It then flows and can be pumped like water. The reactor is also termed a liquid metal reactor or "LMR." Sodium does not slow down neutrons like water does. Further, more neutrons are released in the fission event when the incoming neutron is traveling at a high speed than a low speed. Consequently, there are more neutrons available for capture by U-238 and formation of plutonium. This explains why sodium-cooled reactors can breed more fissionable material than they consume.

It was expected in the 1940s that the first nuclear power reactors would be liquid metal reactors. A small LMR power plant was built in Idaho by Argonne National Laboratory in 1951; it produced enough electricity to light four 200-watt light bulbs. This was the first production of electricity using the energy of nuclear fission. However, the U.S. Navy developed water-cooled reactors for its submarine fleet in the 1950s. These reactors were successful, and American commercial-nuclear-power-plant manufacturers chose to base their designs on water cooling rather than sodium cooling.

Several experimental liquid-metal-reactor power plants have been built and operated since the 1950s by England, France, India, Japan, Russia, and the United States. China recently announced plans to build an LMR. These plants have had safe and generally successful histories, with the following exceptions: the English and the second Japanese units have had sodium-leak problems outside the reactors, and a large French LMR has had design problems inside the reactor. A major purpose of building experimental units of any machines (reactors, airplanes, automobiles) is to uncover unexpected problems. Those encountered here should not be surprising; all should be correctable with reasonable engineering effort.

Officials under President Clinton have chosen to stop breeder reactor development in the United States. However, other nations such as Japan are actively continuing with their development. Russia has breeder reactor power plants; it apparently intends to build more. It also expects to use the breeder plants to desalinate water — to purify ocean or other salty water and make it suitable for uses such as irrigation.

There seems little doubt that breeders will be available to extend our natural uranium supplies.

Diversion of Weapons Material from Breeders

It was noted earlier that officials in Presidents Ford's and Carter's administrations worried about the existence of pure plutonium if spent fuel is chemically reprocessed; a similar situation exists in President Clinton's administration. This has been the focus of their concern about breeder reactors because all existing breeders use pure plutonium. This is because the fuel rods must be discharged periodically and chemically reprocessed. At that point, pure plutonium is recovered, made into new fuel elements, and recharged into the reactor.

The United States has recently developed an improved breeder reactor at Argonne National Laboratory that avoids that problem. It is called the Integral Fast Reactor or IFR. A different type of chemical reprocessing system and a different type of fuel rod are used. Pure plutonium never exists. The plutonium always remains mixed with fission fragments and transuranic elements. This mixture is always intensely radioactive, and diversion would be very difficult. Even if material were stolen, a bomb could not be made without further chemical separation; as noted earlier, highly pure plutonium is needed for such a device. Thus, this new technology should largely eliminate concerns about breeders.

The components of a small-scale power plant using an improved breeder reactor were assembled in Idaho in 1995; the purpose was to demonstrate the capability of a complete system. This small-scale demonstration had cost many tens of millions of dollars and required many year's of effort to assemble. However, Congress, at the urging of the Clinton administration, chose to terminate the program for budgetary and other reasons. This action was taken even though Japanese and private industry contributions would have reduced the remaining demonstration cost to a very minor amount. I believe this action was misguided and unfortunate.

Chapter 10

NUCLEAR POWER COSTS

The cost of electricity from nuclear energy must be competitive with that from other energy sources; otherwise, it will not be used. This will be especially true in the coming years in which the utility industry will be deregulated. Utilities will no longer be monopolies, and competition will reign. Everyone wants to pay as little as possible for electricity.

This book is about the outlook for the future. Therefore, we want to explore the cost of producing electricity in new plants built to meet our future needs.

It is not simple to calculate the cost of making electricity in a new plant. Ten years can pass between the initial decision to build the plant and the time it goes into operation; most plants will then operate for 30 years to as long as 60 years. This 40-year or so time span complicates matters. Let us look at the factors that influence the cost of electricity from nuclear, coal, and natural gas plants; nuclear power competes primarily with electricity from the latter two energy sources.

Factors Influencing Costs

The cost to make electricity in a new plant will depend on several factors including the following:

- The cost to build the plant. Nuclear plants are relatively complex, require expensive materials, and are relatively expensive to build. Coal plants will have similar costs, while gas plants are simpler and less expensive.

- The interest the owner has to pay on the money he or she borrows to build the plant. He or she is like a person who takes out a mortgage to build a home and then pays off the loan with interest over several years.

  These building and interest costs or "capital costs" are termed "fixed costs"; they remain constant and must be paid whether the plant operates or not.

- The cost to operate and maintain the plant including the cost of the fuel (uranium, coal, gas). This is called the "production cost." Average production costs for electric utilities have varied recently, as shown in Fig. 14.

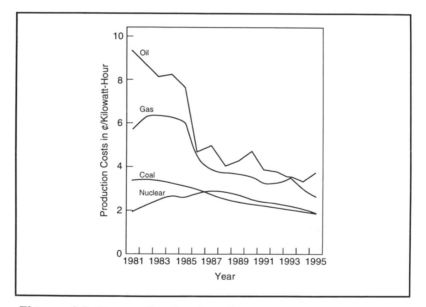

Figure 14: Power Production Costs

Some production costs vary with how the plant is operated; for example, if it is shut down, there are no fuel costs. Others are fixed; the plant staff must be paid whether the plant operates or not.

- The amount of time the plant operates daily. If it operates near 100% capacity 24 hours per day, we call it a "base-load" plant. If it operates only briefly for peak loads, such as for air-conditioning on hot summer days, it is called a "peaking" plant.

Uranium and coal plants are good for base-load plants; their large fixed costs can be spread over a large amount of electricity. The fixed cost per kilowatt-hour will be low.

In contrast, gas plants are good for peakers; with low fixed charges, they do not need to produce large amounts of electricity to hold the cost per kilowatt-hour down. Production costs (especially fuel costs) are higher, but this does not matter if they are not operated much.

- The life of the plant. The entire construction cost, including interest, must be recovered from the electricity sold. If the plant has a high capital cost, a short life, and generates relatively little electricity, the price for each kilowatt-hour of electricity sold must be high to recover the cost.

Note in Fig. 14 how production costs have declined with time; competition has probably caused this. Further changes (down *or* up) will surely occur during the next 40 years. In considering the cost of electricity from a new plant, one must look at how costs are expected to change from year to year throughout the life of the plant.

The cost to build and operate nuclear power plants in the 1960s and 1970s was relatively low; nuclear power was considered cheap. The situation changed in the 1980s. Interest rates were high. Long delays in construction were suddenly encountered, frequently because of legal suits. The Nuclear Regulatory Commission required that changes be made after construction had begun at many plants because of the Three Mile Island accident. Operating costs rose, in part to meet new Commission requirements. The net result was that nuclear power from many plants built in the 1980s or early 1990s became costly. Construction costs were sometimes three or four times greater than planned.

Cost of Electricity from New Plants

As discussed earlier, new power plants will be needed soon; our population and economy have both continued to grow. An important question becomes: How will the cost of electricity from new nuclear, coal, and gas plants compare over the life of the plants?

If fuel costs remained constant at today's prices, electricity from natural gas would probably be 10% or 15% cheaper than from uranium or coal. Although gas production costs are high, low construction costs more than compensate.   However, we may run low on natural gas supplies as we use more and more of it. Difficulty is also encountered in building gas pipelines. These two factors lead many people to believe that gas prices will rise considerably faster than either coal or uranium prices.

If gas prices do rise faster, Electric Power Research Institute studies indicate that nuclear plants would likely generate electricity at a cost no more than 5% to 10% above those for gas and coal plants. These studies assume that standardized plants would be built under the improved Nuclear Regulatory Commission-licensing procedure discussed earlier. Thus, nuclear power is close to being competitive with electricity from both coal and gas plants.

Further, the United States has agreed with other nations that we must all decrease the amount of CO_2 we discharge into the atmosphere.   Congress is considering a tax on all carbon emissions as one means to meet our obligation; the tax would discourage burning fossil fuels. Such a tax could increase the cost for coal or gas electricity by several percent, depending on the size of the tax.

Decommissioning Costs

When nuclear plants are no longer usable, they must be disposed of or "decommissioned" so that they are not hazardous to public health. This is a component of nuclear-electricity cost that does not apply to fossil-fuel electricity.

The money for decommissioning is collected from customers as the plant is operated as part of the price of electricity. Utilities are presently collecting between one-tenth and two-tenths of a cent per kilowatt-hour to fund decommissioning. The estimated cost of nuclear electricity from the new plants discussed above includes funds for decommissioning. A little over 2% of the estimated cost is for decommissioning.

Ordering New Nuclear Plants

A decision by an industrial executive to build a nuclear plant will depend on more than the previously mentioned cost comparisons. Many uncertainties are involved. Nuclear electricity will likely need to be significantly cheaper than that from coal or gas before new plants are ordered in the United States. This is partly because of the cost overruns of the 1980s plants. Investors may feel that recovering their investments from nuclear plants is risky; they may require a greater rate of return to compensate for the risk. Reasonable resolution of the high-level waste issues will almost certainly be required. Nuclear plants are also big and costly; a company may be concerned about investing too much of its resources in one plant or one technology — "about putting too many of its eggs in one basket." In addition, public utility organizations may be hesitant to build nuclear plants if public officials or antinuclear groups oppose.

There are also uncertainties with coal and gas plants, including fuel costs and the carbon tax, discussed previously. Restrictions on the release of sulfur and nitrogen compounds that are responsible for much acid rain could be costly; many lakes in the Adirondack Mountains are purported to be void of fish and most life because of sulfur dioxide emissions. Finally, deaths to the public caused by coal-plant emissions (as discussed earlier) introduce large uncertainty.

Summary

It appears likely that new nuclear plants will produce electricity almost as cheaply as its competitors or even more cheaply if environmental factors enter decisively. Decisions on whether to build such plants will be based on the judgment of prospective owners; they will need to weigh a wide variety of factors. Any decision will be uncertain and risky to some extent.

Other Countries

Decisions on what kind of power plants to build are easier to reach in many other countries — especially those having no coal or natural gas supplies. Nuclear electricity is highly competitive in parts of Europe; France exports large quantities for a profit. A Canadian utility profitably exports electricity to the United States. South Korean officials have described nuclear electricity as their cheapest form. Japan and Taiwan are actively building nuclear plants. China has plans to build over 100 large plants in the next 25 years.

Chapter 11

THE PROMISES OF NUCLEAR POWER

The use of nuclear energy to generate electricity promises great benefits to you as an individual, to the nation, and to the world. These include the following.

1. *Clean Air:* Of all practical means for generating large amounts of electricity, nuclear power is the least harmful to the environment. It emits no CO_2 to cause the greenhouse effect as do coal and natural gas. It emits no sulfur compounds to cause acid rain as does coal nor nitrogen compounds as do both coal and natural gas. Nuclear power plants cause no silting of pristine river systems and no large loss of farms, homes, and wilderness to reservoirs such as do hydroelectric plants. A strong argument can be made that nuclear power has no significant harmful effect on the environment at all. In 1994, America's 109 nuclear plants spared the atmosphere about 500 million tons of CO_2, five million tons of sulfur dioxide, and 2.5 million tons of nitrogen oxides. It is obvious that substitution of nuclear electricity for fossil fuel electricity would result in a cleaner atmosphere.

2. *Resource Conservation:* Coal, petroleum, and natural gas represent precious natural resources built up over millions of years. They have many uses, such as feed stocks for medicines, plastics, and other industrial products, and we should not squander them when substitutes are available. Uranium, in contrast, has no use except for power production, atomic weapons, and a few minor applications such as to serve as ballast in ships.

3. *Saving Lives:* As discussed previously, nuclear power has been demonstrated to be safer than power from coal. The evidence indicates more people lose their lives in the United States each year from coal electricity than will lose their lives worldwide over the next 60 or 70 years as a result of the Chernobyl accident.

4. *Preventing Wars:* Along with battery-driven or hydrogen-driven automobiles, nuclear power has the potential to prevent world wars over Middle East oil supplies. Middle East oil is crucial to the well-being of many nations as long as they rely so heavily on gasoline-driven automobiles; the United States led the Gulf War in 1991 (at a cost of tens of billions of dollars) to prevent loss of control of that oil. The Middle East currently supplies 30% of the world's oil; this will likely rise to 50% within a few years. Development of practical, inexpensive storage batteries for electric cars, coupled with nuclear electricity to charge the batteries, could greatly reduce our need for gasoline.

    Development of hydrogen-driven automobiles would accomplish the same objective. In this case, nuclear electricity would be used to separate the hydrogen from the oxygen in water. Hydrogen is not normally available as a separate material. The hydrogen would then be recombined with oxygen in a fuel cell to release energy and propel the car.

The financial cost of preparing for and fighting wars is very high; we could gain by subsidizing heavily the cost of battery or hydrogen-driven autos. The human cost of fighting wars is also very high; philosophically, it should be unacceptable in modern society.

5. *Improving Our Economy:* The use of battery or hydrogen-driven autos would also drastically decrease the cost of importing oil. At the present time, we spend about $50 billion per year as a nation to import petroleum; some economists predict this will climb to $100 billion per year in a few years. Many economists believe these expenditures cause a serious drain on our economy — that they cost many jobs and lower our standard of living.

6. *Further Improvement of Our Air Quality:* Gasoline is a major contributor to smog formation and air pollution. Substitution of battery or hydrogen-driven autos for gasoline-driven vehicles would obviously have a great impact on air quality. For example, in the fuel cell discussed in item 4, the waste from the cell would be ordinary water — when the hydrogen and oxygen combined, water would be formed again. It is difficult to envision a more environmentally friendly system.

7. *An Almost-Unlimited Power Supply:* It is vital that our nation (and every nation) have an ensured, long-range supply of electricity; there is enough uranium underground and in the oceans to meet our electricity needs for centuries using advanced reactors.

8. *Lower Electricity Costs:* Newer plants will be less expensive to build and operate than the present first-generation units. Further, large additional savings in electricity costs could come if the question of whether low level radiation is harmful could be resolved. If it is not harmful, current power plant design, construction, operation, and management are grossly conservative; electricity costs could

be reduced substantially. As a minimum, costs could be reduced considerably if the public were simply less fearful of radiation.

In the 1960s and 1970s before nuclear power costs climbed, people dreamed of many benefits of cheap electricity. One dream was that it might make possible the desalination of water at affordable prices — that low-cost water recovered from the oceans could be used to irrigate the world's deserts and grow food. This dream still exists — that we could *make the deserts bloom.*

Chapter 12

WHAT CAN WE DO?

This author believes that the expanded use of nuclear power around the world is inevitable as the need for electricity increases. Many countries have neither coal nor natural gas; others do not have rail systems to transport their coal to population centers where electricity is needed. Because uranium is easy to transport and stockpile, many countries will rely on it for a secure energy supply. They will likely have no other practical choice. All countries also face the greenhouse problem.

I also believe that there will be a resurgence in building nuclear power plants in the United States. This will happen (a) as our need for electricity grows, and (b) when our electricity-generating companies conclude that nuclear is their best energy choice. Companies will reach that conclusion when some combination of factors such as the following exists:

- When the economic cost is clearly favorable for nuclear power.

- When our national energy policy places increased emphasis on clean air.

- When that policy gives greater recognition to saving lives.

- When the policy places greater emphasis on freeing the nation from the political turbulence associated with Middle East oil.

- If and when Congress and the President develop a long-range energy policy agreed upon by both political parties.

- When a satisfactory HLW storage system is established.

- When nuclear power "becomes popular again" — when there is recognition of the costs of *not* using nuclear power; when it is not fashionable for environmental groups to oppose nuclear energy; when political leaders or candidates can not use an antinuclear platform to gain votes. The general public already looks favorably on the use of nuclear power. The formal position of Congress is to develop the use of nuclear energy for the benefit of society.

However, simply waiting for the inevitable to occur is not good enough. We can hasten the resurgence of nuclear power and bring about some of the immense benefits promised by it. Among the things we can do are the following.

First, you must weigh the facts and conclude that nuclear power does indeed offer great benefits. Then, we all must be willing to speak out in its favor.

We must challenge environmental and other groups that oppose nuclear power — to insist that they examine the facts and act accordingly. One effective way to do so is to stop supporting them financially if they resist.

We must speak out when the mass media — newspapers and TV in particular — make erroneous statements. We must oppose their use of dramatic headlines and stories that mislead and scare the public.

It is especially important to make our views known to our political leaders. They must be made to recognize that nuclear power gives us clean air today and saves lives today — and that we could be dumping far fewer pollutants into the air and saving thousands more lives. Letters to both the President and Vice President and to our Senators and Representatives are needed. Letters to state political leaders are also necessary. Public Service Commissions should be urged to compare not only economic but also health and environmental costs of different fuels.

We should urge Congress and the President to continue to support the development and use of nuclear power for the benefit of all of society.

In closing, we must be aggressive to gain the enormous benefits that nuclear power offers. Recent history has demonstrated that those benefits will not come easily.

DEFINITION OF TERMS AND ABBREVIATIONS

Acute Doses of Radiation: Doses received in a short-time period such as minutes or hours.

CO_2: Carbon dioxide.

HLW: High-level waste, the material in used or spent fuel rods. See Chapter 7.

Isotope: Any of two or more atoms of an element which have the same number of protons but different numbers of neutrons in the nucleus. See Chapter 2.

Linear, No-Threshold Theory: The theory that radiation presents a health risk that is proportional to dose, no matter how small the dose. See Chapter 5.

Micron: One micron is equal to 1/70 the width of a human hair or 1/25,000 of an inch.

NRC: U.S. Nuclear Regulatory Commission.

One Year of Background Radiation: The amount of radiation the average American absorbs from natural sources each year. This radiation comes from four main sources as follows: (The amounts are given in standard terms called milliSieverts and millirems.)

Origin	milliSieverts	millirems
From the sun and outer space	0.27	27
From the earth, including uranium and thorium	0.28	28
From inside our bodies, including potassium	0.39	39
Radon (from our buildings and the ground)	2.0	200
Total	3.0	300

B. G. Bennett of the United Nations Scientific Committee on the Effects of Atomic Radiation gives the average radiation dose to the world's population from natural radiation sources as 2.4 milliSieverts per year.

Particulates: Particulates are composed of (a) solid particles emitted to the atmosphere such as dust or the soot from power plants or wood stoves, and (b) other solid particles or liquid droplets formed in the atmosphere. The latter are called aerosols, and some are formed from the sulfur oxides and nitrogen oxides emitted when fossil fuels are burned. These fossil-fuel aerosols are usually smaller than one micron in diameter. Scientists believe the most harm comes from particles smaller than one micron.

Suggested Reading

General

- *Bluebells and Nuclear Energy* by Albert B. Reynolds, Medical Physics Publishing, 4513 Vernon Blvd., Madison, WI 53705, 1996.

- *The Nuclear Energy Option: An Alternative for the 90s* by Bernard L. Cohen, Plenum Press, New York, 1990.

- *America The Powerless — Facing our Nuclear Energy Dilemma* by Alan Waltar, Medical Physics Publishing, 4513 Vernon Blvd., Madison, WI 53705, 1995.

Chapter 5

- *The Good News About Radiation* by John Lenihan, Medical Physics Publishing, 4513 Vernon Blvd., Madison, WI 53705, 1993.

- *Understanding Radiation* by Bjorn Wahlstrom, Medical Physics Publishing, 4513 Vernon Blvd., Madison, WI 53705, 1995.

- *Health Effects of Low-Level Radiation* by Sohei Kondo, Medical Physics Publishing, 4513 Vernon Blvd., Madison, WI 53705, 1993.

- *Power Production: What Are the Risks?* by J. H. Fremlin, Adam Hilger, New York, 1989.

- "Natural Background Radiation Exposures World-Wide," B. G. Bennett, International Conference on High Levels of Natural Radiation, Ramsar, Iran, 3-7 November 1990, INIS-mf-13747.

- "Studies of the Mortality of Atomic Bomb Survivors." Report 12, Part 1. Cancer: 1950-1990 by Pierce, Shimizu, Preston, Vaeth, and Mabuchi, *Radiation Research*, Vol. 146, July 1996, pages 1-27.

- "The Children of Parents Exposed to Atomic Bombs: Estimates of the Genetic Doubling Dose of Radiation for Humans," Neel, Schull, Awa, Satoh, Kato, Otake, and Yoshimoto, *Am. J. Hum. Genet.*, Vol. 46, 1990, pages 1053-1072.

- "Sources and Effects of Ionizing Radiation," United Nations Scientific Committee on the Effects of Atomic Radiation, New York, 1993.

- "Health Effects of Exposure to Low Levels of Ionizing Radiation - BEIR V," Committee on the Biological Effects of Ionizing Radiation, National Academy Press, 1990.

- "Beneficial Radiation," Zbigniew Jaworowski, *Nukleonika*, Vol. 40, No. 1, 1995, pages 3-12.

- "Chernobyl — Ten Years On: Radiological and Health Impact," Nuclear Energy Agency, Organisation for Economic Co-Operation and Development, Paris, November 1995.

- "One Decade After Chernobyl: Summing Up the Consequences," Kaul, Landfermann, and Thieme, *Health Physics*, Vol. 71, No. 5, Nov. 1996, pages 634-640.

Chapter 6

- "Breath Taking — Premature Mortality Due to Particulate Air Pollution in 239 American Cities," Principal Author Deborah Sheiman Shprentz, Natural Resources Defense Council, 40 West 20th Street, New York, NY 10011, May 1996.

Chapter 7

- "Plutonium: Facts and Inferences," Electric Power Research Institute, Palo Alto, CA, EPRI EA-43-SR, August 1976.

- "Radioactive Waste Transportation — A Guide for Midwestern Legislators," The Council of State Governments, Midwestern Office, 641 E. Butterfield Road, Suite 401, Lombard, IL 60148-5651, DOE/RW/00286-10, December 1996.

- "The State Role in Spent Fuel Transportation Safety," National Conference of State Legislatures, 1560 Broadway, Denver, CO, May 1996.

Chapter 8

- "Reactor-Grade Plutonium's Explosive Properties," J. Carson Mark, Nuclear Control Institute, Washington, D.C., August 1990.

- "A Perspective on the Dangers of Plutonium," Sutcliffe, Condit, Mansfield, Myers, Layton, and Murphy, Center for Security and Technology Studies, Lawrence Livermore National Laboratory, UCRL-ID-118825, April 14, 1995.

- *Adventures of a Physicist* by Luis W. Alvarez, Basic Books Inc., New York, 1987.

Chapter 10

- "Ensuring the Economic Competitiveness of Advanced Light Water Reactors," Santucci and Sliter, SFENS/ENS Topical Meeting TOPNUX '96, Paris, September 30, 1996.,